Biosensors

Biosensors

An Introductory Textbook

edited by

Jagriti Narang
Chandra Shekhar Pundir

PAN STANFORD PUBLISHING

Published by

Pan Stanford Publishing Pte. Ltd.
Penthouse Level, Suntec Tower 3
8 Temasek Boulevard
Singapore 038988

Email: editorial@panstanford.com
Web: www.panstanford.com

British Library Cataloguing-in-Publication Data
A catalogue record for this book is available from the British Library.

ISBN 978-981-4745-94-9 (Hardcover)
ISBN 978-1-315-15652-1 (eBook)

Printed in Great Britain by Ashford Colour Press Ltd

Contents

3. Characterization Techniques   **43**

Jagriti Narang, Nitesh Malhotra, and Rachna Rawal

Jagriti Narang, Nitesh Malhotra, and
Chandra Shekhar Pundir

Preface

Nanomaterials play an important role in sensing due to their fascinating features such as large surface area, biocompatibility, and promotion of electron transfer kinetics. As implied by its title, *Biosensors: An Introductory Textbook*, this book discusses the preparation as well as the characterization of nanomaterials and their application in biosensors, with equal emphasis on both. It predominantly demonstrates the construction of all types of biosensors from scratch. A biosensor is an analytical device that combines a transducer with a biologically sensitive and selective component. When a specific target molecule interacts with the biological component, a signal is produced at transducer level, which is proportional to the concentration of the substance. This characteristic enables biosensors to measure the concentrations of various compounds present in an environment, chemical processes, food, and human body at a much lower cost compared with traditional analytical techniques.

This book focuses on the basic concepts of nanomaterials and their application in sensing. The first two chapters discuss the methods for preparation of various nanomaterials. The third chapter encompasses techniques that can be used to characterize the nanomaterials and fabricate different sensors. The next three chapters are devoted to the techniques involved in the fabrication of biosensors and the applications of biosensors in different fields of diagnostics. The last chapter highlights a new area in sensing research, which is fabrication and application of microfluidic devices. This book focuses on the unique electrical, optical, mechanical, thermal, and vibrational properties of nanomaterials. It also discusses the various aspects of how and where these structures can be applied with enhanced properties, with evidence from published literature. It can be a great resource for budding nanotechnology researchers, undergraduates, and post-graduate students who are interested in perusing science and embarking on a career in sensing research, which has been one of the most-explored fields of research and has played a major role in the current diagnostic boom.

We express our appreciation and gratitude to all authors who contributed their research results to this book and to the team at Pan Stanford Publishing that worked with professionalism and dedication to bring it out.

Jagriti Narang
Chandra Shekhar Pundir
Spring 2017

Acknowledgment

We would like to thank all scientists referenced throughout our book whose work greatly helped in the compilation of information related to the topic.

Dr. Narang would like to thank her parents who always encourage her to accomplish various new tasks. She also dedicates this book to all working women who keep their children out of sight with heavy hearts.

Chapter 1

Nanomaterials

Jagriti Narang,[a] Nitesh Malhotra,[b] Chandra Shekhar Pundir,[c] and Tulika Dhaiya[d]

[a]*Amity Institute of Nanotechnology, Amity University, Noida-201313, India*
[b]*Amity Institute of Physiotherapy, Amity University, Noida-201313, India*
[c]*Department of Biochemistry, MD University, Rohtak-124001, India*
[d]*Department of Biotechnology, Shoolini University, Solan-173229, India*
jags_biotech@yahoo.co.in

1.1 Introduction

Nanotechnology is a wide and interdisciplinary area of research and development activity that has been rising worldwide since the last decade. Nanomaterials are the foundation of nanotechnology. Nanomaterials are materials that are characterized in the size range of 1–100 nm. A nanometer is one millionth of a millimeter, approximately 100,000 times smaller than the diameter of a human hair [1–4].

1.2 Occurrence of Nanomaterials

Some nanomaterials occur naturally. These materials are associated with the natural world (animal and mineral) without any engineering

Biosensors: An Introductory Textbook
Edited by Jagriti Narang and Chandra Shekhar Pundir
Copyright © 2017 Pan Stanford Publishing Pte. Ltd.
ISBN 978-981-4745-94-9 (Hardcover), 978-1-315-15652-1 (eBook)
www.panstanford.com

by human beings, such as nanoparticles, and evolved from natural erosion, volcanic activity, and clays. Nanoparticles are also dispersed in natural colloids such as milk and blood. By learning and gathering motivation from nature, biomaterials have been designed and engineered for many applications [5]. Engineered nanomaterials are already being used in many commercial products and processes. Nanoparticles can be found in sunscreens, cosmetics, sporting goods, stain-resistant clothing, tires, electronics, as well as in day-to-day activities. Nanomaterials are also employed in nanomedicine application such as for diagnosis, imaging, and drug delivery.

Nanomaterials have remarkable distinct properties that are normally not observed in their bulk counterparts. There are two most important properties that make them best in comparison to their bulk counterparts: large surface area and novel quantum effects [6–9]. An additional key benefit of nanomaterials is the enhancement of their fundamental properties such as magnetization, optical properties, melting point, and hardness [10–15] as compared to counter bulk materials without change in chemical composition.

1.3   Revolution in Nanomaterials

Nature has evolved many nanostructures such as skin, claws, beaks, feather, horns, spider silk, and lotus leaf. Nanoscaled smoke particles were formed during the use of fire by early humans. The scientific story of nanomaterials, however, began much later. One of the first scientific reports is the colloidal gold particles synthesized by Michael Faraday as early as 1857. Nanostructured catalysts have also been investigated for more than 70 years. By the early 1940s, precipitated and fumed silica nanoparticles were manufactured and sold in the USA and Germany as substitutes for ultrafine carbon black for rubber reinforcements [16, 17].

1.4   Classification of Nanomaterials

Nanomaterials have extremely small size, having at least one dimension of 100 nm or less. They are classified based on the number of dimensions, which are not confined to the nanoscale range (<100 nm): (i) zero-dimensional (0-D), (ii) one-dimensional (1-D), (iii) two-dimensional (2-D), and (iv) three-dimensional (3-D) [18](Fig. 1.1).

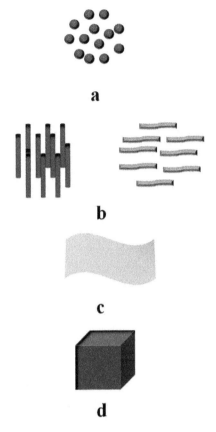

Figure 1.1 Classification of nanomaterials: (a) 0-D spheres and clusters; (b) 1-D nanofibers, wires, and rods; (c) 2-D films, plates, and networks; (d) 3-D nanomaterials.

1.4.1 Zero-Dimensional

In 0-D materials, all dimensions are measured within the nanoscale (no dimensions are larger than 100 nm). Characteristic features of 0-D nanomaterial are as follows:

- They can be amorphous or crystalline.
- They can be single crystalline or polycrystalline.
- They can possess different shapes and forms such as nanotubes, dendrimers, and fullerenes.
- They can be metallic, ceramic, or polymeric.

1.4.2 One-Dimensional

In 1-D materials, two dimensions are restricted to the nanoscale and one dimension is restricted to the macroscale. One dimension leads to needle-shaped nanomaterials. Characteristic features of 1-D nanomaterials are as follows:

- They include nanotubes, nanorods, and nanowires.
- They can be amorphous or crystalline.
- They can be chemically pure or impure.
- They can be metallic, ceramic, or polymeric.

1.4.3 Two-Dimensional

In 2-D materials, one dimension is restricted to nanoscale and two dimensions are restricted to the macroscale. Two dimensions lead to plate-like shapes. Characteristic features of 2-D nanomaterials are as follows:

- These materials include nanofilms, nanolayers, and nanocoatings.
- They can be amorphous or crystalline.
- They can be made up of various chemical compositions.
- They can be deposited on a substrate.
- They can be embedded into the surrounding matrix material.
- They can be metallic, ceramic, or polymeric.

1.4.4 Three-Dimensional

In 3-D materials, none of the dimensions are confined to the nanoscale. Three dimensions lead to nanocrystalline structures. Characteristic features of 3-D nanomaterials are as follows:

- These materials can be dispersions of nanoparticles, bundles of nanowires, and nanotubes as well as multinanolayers.
- In bulk, they can be composed of a multiple arrangement of nanosized crystals with different orientations.

1.5   Importance of Nanomaterials

The principal advantage of nanoparticles is their size regime, i.e., their large surface-area-to-volume ratio. This feature of nanoparticles enables high surface reactivity with the surrounding surface, which is ideal for many applications such as photocatalysis or in the fabrication of sensing devices [19]. Nanomaterials also show the capability of varying their elementary properties such as magnetization, optical properties, and melting point relative to bulk materials with no change in their chemical composition. These materials have attracted great attention in recent years as they possess mechanical, electrical, optical, and magnetic properties: Ceramics with integrated nanomaterials are of meticulous concern because they show more ductility at elevated temperatures compared to the coarse-grained ceramics; nano-metallic powders have been used for the production of dense parts and porous coating; nanostructured metal clusters are involved in catalytic applications; nanostructured metal-oxide thin films are employed for gas sensors; and metal-oxide nanoparticles are employed for biosensing applications [20, 21]. Nanocrystalline structures are being employed in solar cells.

1.6   Synthesis and Processing of Nanomaterials

Synthesis of nanomaterials with stringent control over size, shape, and crystalline structure has become very important for the applications of nanotechnology in numerous applications such as catalysis, medicine, and electronics. Nanoparticles are basically synthesized through two approaches: top down and bottom up. In the top-down approach, the bulk solid is dissembled into smaller and smaller portions so that the resulting material comes in nanometer. The bottom-up approach involves assembling of atoms in solution to form the material in the nanometer range [22].

1.6.1   Synthesis of Metallic Nanoparticles

Various methods are available for the synthesis of metallic nanoparticles:

1.6.1.1 Sol–gel synthesis

The sol–gel process is a type of wet chemical process. This technique involves the conversion of a system from a colloidal liquid (sol) into a gelation of sol, i.e., a semisolid phase material (gel) [23]. The precursor is dispersed, which forms a metallic oxide upon contact with water or a dilute acid or any type of solution. The second step involves the extraction of the liquid from the dispersed solution, yielding the gel. Calcination of the gel produces the oxide [24–26].

The process in which the precursor is dissolved in solution is called **hydrolysis**, while the process involving the removal of water from the dispersed solution is called **condensation**. The process in which polycondensation occurs by dehydration, forming a viscous solution, is called **gelation**. If dehydration is done by thermal evaporation, the resulting monolith is termed **xerogel**. If dehydration is done under supercritical conditions, the resulting product is an **aerogel**.

The reactions involved in the formation of metal-oxide nanoparticles are as follows:

$$MOR + H_2O \rightarrow MOH + ROH \text{ (hydrolysis)}$$

$$MOH + ROM \rightarrow MOM + ROH \text{ (condensation)}$$

where, MOR and ROM are metal alkoxides, MOH is metal hydroxide, and MOM is metal-oxo metal bridge.

The sol–gel process can be characterized by a series of distinct steps (Fig. 1.2).

Advantages: High purity, isotropic, tunable porosity, and composition and low-temperature annealing.

Disadvantages: Agglomeration, low production.

Example: Iron nanoparticles can also be prepared by mixing the precursor ($FeCl_3$) solution with tetraethylorthosilicate (TEOS), ethanol, and water. This process is called hydrolysis. After that, a few drops of acid are added, which act as a catalyst. Then condensation starts. After polycondensation or dehydration, the resulting gel is reduced or calcined at a temperature of 400°C. Thus iron nanoparticles are formed.

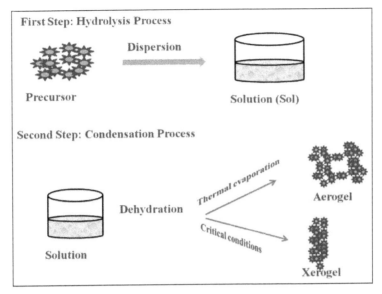

Figure 1.2 Schematic representation of the sol–gel process of synthesis of nanomaterials.

1.6.1.2 Hydrosol method/chemical reduction method

Nanoparticles of noble metals can be prepared by the hydrosol method using a reducing agent [27]. In this method, noble metal salts undergo reduction to produce a colored solution (sol) because of the presence of free electrons on the metal surface in the conduction band and positively charged nuclei. The chemical involves conversion of the dissolved metal precursor to the solid metal by the introduction of a reducing agent. The nature of reducing agents is an important factor in controlling the size and shape distribution of noble metal salt nanoparticles. Hydrosols are striking inorganic materials because of their remarkable applications in photography [28], catalysis [29], biosensors [30], and biomolecular detection [31] and also because of their antimicrobial activities [32–34] and environmentally benign nature [35–38]. In this method, different reducing agents—ascorbic acid [39], hydrazine [40], ammonium formate [41], dimethylformamide [42], and sodium borohydride [43]—are used to produce nanoparticles.

Advantages: Narrow size distribution, no aggregation.

Disadvantages: The nanoparticle surface can absorb the oxidation products of the reducing agent, thus affecting further growth.

Example: Noble metal nanoparticles are prepared by mixing an optimized amount of an aqueous solution of gold or silver salts and reducing agents at room temperature. Gold nanoparticles are produced by the reduction of chloroauric acid with sodium citrate in an aqueous solution on boiling, and silver nanoparticles are produced by the reduction of silver salts with sodium borohydride.

1.6.1.3 Vacuum deposition method

In the vacuum deposition process, elements, alloys, or compounds are vaporized and deposited on the substrate in a vacuum. In this process, the precursor is heated by thermal processes such as arc heating and electron beam heating, and then the substrate evaporates and enters into an inert atmosphere. In this atmosphere, the substrate temperature is lowered by collision with the cold gas, which undergoes condensation into small clusters. The clusters continue to grow, if they remain in the supersaturated region. The size can be controlled depending on the gas pressure, evaporation speed, type of gas used for collision, and substrate temperature.

Advantages: Easy to produce, inherent simplicity, high deposition rates.

Disadvantages: Due to the oxidative atmosphere in the flame, the process is restricted only to the formation of oxides.

Examples: Icosahedral gold nanoparticles are generated from an inert gas aggregation source using helium and deposited on a substarte. Zirconium oxide nanoparticles are also prepared by this method.

1.6.2 Synthesis of Nanotubes and Nanowires

Nanotubes and nanowires turn out to be one of the most promising materials in nanotechnology, because of their distinct properties such as rigidity, high mechanical strength, ductility, and electric conductivity. Two methods are used for the synthesis of nanotubes and nanowires: (i) chemical vapor deposition (CVD) and (ii) chemical condensation processing.

1.6.2.1 Chemical vapor deposition

The main feature of CVD is the deposition that takes place due to a chemical reaction between some reactants on the substrate. In this process, precursors are pumped into a reaction chamber under optimum experimental conditions. One of the products of the reaction gets deposited on the substrate by pumping out the other nonspecific products.

Parameters that affect the deposition are rate of reaction, transport of gaseous materials, and diffusion. In thermal CVD, the reaction is activated by a high temperature, i.e., above 900°C [44–47].

Advantages: Excellent throwing power, production of coatings of uniform thickness and properties, low porosity, selective deposition.

Disadvantages: Sensitive to contamination.

Example: SiC/Si$_3$N composite powder is prepared using CH$_4$ as a source of gas at a high temperature above 1000°C.

1.6.2.2 Chemical vapor condensation

In chemical vapor condensation, vapor phase precursors are pumped into a hot-wall reactor under optimum conditions that favor nucleation of particles in the vapor phase rather than deposition of a film on the wall. It involves pyrolysis of metal organic precursors by the hot-wall reactor in a reduced atmospheric condition. Parameters that affect the preparation of nanoparticles are the decomposition temperature of precursors, temperature, and carrier gases. Moreover, this process can be applied to the formation of doped nanoparticles since in this process, two precursors are supplied at the front end of the reactor and at the second stage of the reactor. Nanoparticles that are formed by homogeneous nucleation are coated by heterogeneous nucleation in the second stage of the reactor [48–52].

Advantages: High rate of production, little agglomeration.

Disadvantages: Sensitive to contamination, oxidation.

Example: Iron nanoparticles are prepared by this method using argon or helium gas as a carrier gas, and iron pentacarbonyl (Fe(CO)$_5$) is taken as the precursor. The precursor decomposes in that furnace, when heat is supplied through the tubular furnace and the carrier gas is passed. These decomposed particles condense into clusters or particles. Particles are deposited on the surface of a liquid nitrogen–cooled chiller in the work chamber, from which powders can be scrapped off and collected (Figs. 1.3 and 1.4).

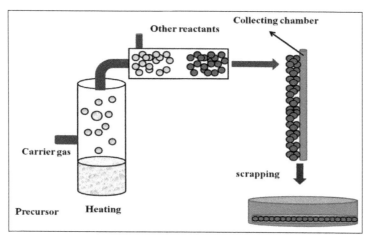

Figure 1.3 Schematic representation of chemical vapor condensation reactor.

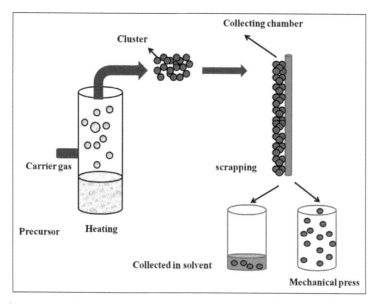

Figure 1.4 Schematic representation of typical set-up for gas condensation synthesis of nanomaterials.

1.7 Biological/Green Synthesis of Nanoparticles

Living organisms, such as microorganisms, and plant extracts are involved in the synthesis of many nanoparticles [53–56]. There are two methods involved in the synthesis of nanoparticles: (i) intracellular and (ii) extracellular synthesis. In the synthesis of nanoparticles, there is a requirement of a precursor of the respective metal and reducing agents. The reducing agent plays an important role in the synthesis of nanomaterials by the reduction process. By choosing different reductants, researchers can produce the desired nanostructures. In the biological synthesis of nanomaterials, reducing agents or stabilizers are utilized from living organisms. Biologically synthesized nanoparticles can be categorized into two groups: (i) bioreduction and (ii) biosorption. Bioreduction involves the reduction of the precursor ions by exudates from biological compounds [57]. Biosorption involves the passive binding of precursor ions from an aqueous solution onto the organism. Certain microorganisms express peptides that bind to the metal ions present in the soil solution, and these can form stable complexes in the form of nanoparticles [58].

Prokaryotes are the choice of researchers, as they are cosmopolitan and easy to cultivate, and their capacity to acclimatize under unfavorable conditions makes them the best candidate for the synthesis of nanoparticles. Growth conditions such as temperature, oxygenation, and incubation time can be easily controlled. Various species of bacteria are exploited to synthesize nanoparticles, as depicted in Table 1.1.

1.7.1 Mechanism of Biological Synthesis of Nanoparticles

The route for the biological synthesis of nanoparticles involves responses of various biological agents, which are different for different metal ions. Microorganisms follow different pathways, either intracellular or extracellular, for the synthesis of nanomaterials. But the mechanism behind intracellular and extracellular synthesis of nanoparticles is different for different biological agents. The intra-

Table 1.1 Microorganisms employed in the green synthesis of various metal nanoparticles

Microorganism	Nanoparticles
Bacillus licheniformis	Silver nanoparticles [59]
Delftia acidovorans	Gold nanoparticles [60]
Rhodopseudomonas capsulate	Gold nanoparticles [61]
Escherichia coli	Pallidium nanoparticles [62]
Lactobacillus spp., *Pediococcus pentosaceus, Enterococcus faecium*, and *Lactococcus garvieae*	Silver nanoparticles [63]
Bacillus sphaericus	Ag and Au nanoparticles [64]
Morganella morganii	Cu nanoparticles [65]
Fusarium oxysporum	Silver nanoparticles [66]

cellular pathway involves the absorption of ions on the surface of the microbial cell because the charge of the cell wall attracts positively charged metal ions. Biosorption occurs through electrostatic interaction. The enzymes present within the cell wall act as a reducing agent and reduce the ions to nanoparticles. After their formation, the nanoparticles diffuse out of the cell wall. Three steps are involved in the biogenic synthesis of nanoparticles: (i) trapping, (ii) bioreduction, and (iii) capping. Metal ions get trapped in the cell wall through electrostatic interaction. The second step involves bioreduction in which metal ions get reduced by enzymes present within the cell wall [67]. Afterward, the small-sized nanoparticles diffuse out of the bacterial cell wall (Fig. 1.5).

The extracellular pathway for the synthesis of nanoparticles involves the synthesis of nanoparticles outside the cell, and the enzymes involved in the synthesis of nanoparticles are present outside the cell wall; for example, the enzymes secreted by the microorganisms facilitate in the bioreduction of metal ions (Fig. 1.6). Various enzymes are involved in the extracellular synthesis of nanoparticles, such as nitrate reductase, for the extracellular synthesis of nanoparticles [68–72].

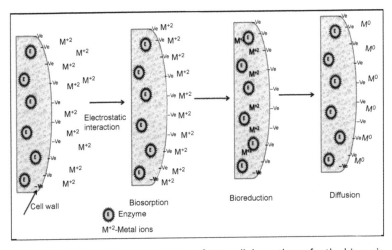

Figure 1.5 Schematic representation of intracellular pathway for the biogenic synthesis of nanoparticles.

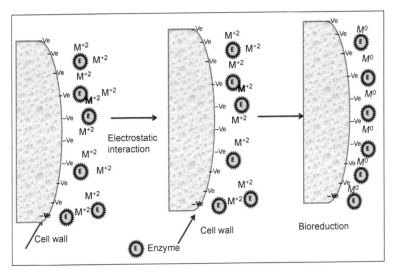

Figure 1.6 Schematic representation of extracellular pathway for the biogenic synthesis of nanoparticles.

1.8 Properties of Nanomaterials

Nanomaterials have structural properties that are in between those of atoms and bulk materials. Materials in the range of micrometers show properties analogous to the corresponding bulk materials, while materials down to nanometer dimensions exhibit considerably different properties from those of bulk materials. This is because nanomaterials possess certain distinct features such as large surface area, high surface energy, and reduced imperfections, which do not exist in the corresponding bulk materials.

Properties of nanomaterial mean that a material shows its best characteristics under certain conditions. Table 1.2 summarizes various properties of nanomaterials with their characters and reasons.

Table 1.2 Various properties of nanomaterials with their characters and reasons

Properties	Characters	Reason
Optical properties	Color, transparency	Nanoparticles are so small that electrons are not free to be in motion. When electron movement is restricted, these particles act in a different way with light.
Electrical properties	Conductivity	Nanoparticles are so small that the electron becomes more confined in the particle, which leads to an increase in the band gap energy. The valence and conduction bands break into quantized energy levels.
Physical properties	Hardness, melting point	The percentage of surface atoms is more in the case of nanoparticles, and surface atoms require less energy to move because they are in contact with fewer atoms of the substance. Therefore, the melting point of nanoparticles is lower than that of bulk materials.
Chemical properties	Reactivity, reaction rates	Nanoparticles show high surface-area-to-volume ratio, which results in better catalysis since a larger fraction of the material is exposed for potential reaction.

1.8.1 Optical Properties

Optical properties show interesting and practical aspects of nanomaterials. The parameters that affect the optical properties of nanomaterials are size, shape, surface characteristics, and aggregation.

- **Shape:** Plasmon resonance depends mainly on the dielectric ambient conditions. This effect is originated by the dependence of the particle polarizability on the dielectric properties of the environment. A change in the polarizability after modification in the environment is determined by the local field surrounding the nanoparticle surface, which depends on the particle shape and size. Therefore, shape affects the plasmon resonance.

- **Size:** If the size of particles is small, then electrons are not free to move as in the corresponding bulk materials. Thus, particles act differentially with light. The scattering of visible light is mainly influenced by particle size and the difference between the refractive index of the pigment and the surrounding media. If the size is small, the particles normally do not scatter. For instance, zinc oxide powder looks white in color, while zinc oxide nanopowder makes transparent dispersions.

- **Surface characteristics:** If the refractive index near the nanoparticle surface increases, the nanoparticle extinction spectrum shifts to longer wavelengths, and if the index decreases, the spectrum shifts to shorter wavelengths.

- **Aggregation:** When particles aggregate, the conduction electrons near each particle surface become delocalized, shifting the absorption and scattering peaks to longer wavelengths. For instance, dispersed silver nanoparticles show yellow color, while aggregated silver nanoparticles show grey color.

Applications: The optical properties have applications in optical detectors, lasers, sensors, imaging, phosphor, display, solar cells, photocatalysis, photoelectrochemistry, and biomedicine.

1.8.2   Electrical Properties

Electrical properties of nanomaterials vary among materials. They mainly depend on the diameter of the nanomaterials. High electrical conductivity of nanomaterials is due to the reduced imperfections in the material. Properties such as conductivity or resistivity come under the category of electrical properties. The conductivity of nanomaterials depends on various parameters such as area of cross section and shear force, giving twist to materials. For instance, the conductivity of a multi-walled carbon nanotube is different from that of a single nanotube of same dimensions.

Applications: The electrical properties have applications in sensors, imaging, and biomedicine.

1.8.3   Mechanical Properties

Materials at the nanoscale level show drastic change in physical properties. For instance, all cutting tools are made up of nanomaterials since these materials are stronger, harder, and with reduced imperfections. Automobile engines are coated with nanocrystalline substances since they maintain heat more efficiently, which results in complete combustion of fuel. These ceramic-based nanomaterials are also exploited in ball bearings, filters, and furnaces.

1.8.4   Chemical Properties

Nanoparticles show high surface-area-to-volume ratio, which is inversely proportional to the size. Due to the small size of nanoparticles, most of the atoms are present on the surface than inside. Due to the surface characteristics, nanoparticles become more chemically active, because most of the chemical reactions occur at the surfaces. For instance, gold remains inert in nature, but in the nanoscale form, it shows reactivity.

1.9   Conclusion and Future Perspective

The scope of nanomaterials is colossal and related to applied sciences. Our knowledge of nanomaterials is still partial, and we have a long way to go. A better understanding of the preparation of nanomaterials will assist in innovation in applied sciences. Various methods are available for the synthesis of nanomaterials, such as physical, chemical, and biological. However, there is enormous scope for the synthesis of nanoparticles with regard to cost. There is a need to develop inexpensive and tuned methods for synthesizing nanomaterials. Materials at the nano level show distinct and astonishing properties other than bulk materials. With the development of new methods that can synthesize nanomaterials at the nanometer scale, it has become increasingly possible to synthesize nanomaterials of desired size. Yet there is a lot to explore in the synthesis of nanomaterials and exploitation of nanomaterials in various applications.

Problems

1. What are nanoparticles, nanotubes, and nanoplates?
2. What are the uses of nanoparticles?
3. Describe the advantages of nanomaterials in biosensor fabrication?
4. What are the major advantages and disadvantages of green synthesis of nanoparticles?
5. What are the advantages of the sol–gel process in the synthesis of nanomaterials?

References

1. Varma, R. S., Saini, R. K., and Dahiya, R. (1997). Active manganese dioxide on silica: Oxidation of alcohols under solvent free conditions using microwave, *Tetrahedron Lett.*, **38**, pp. 7823.

2. Kidwai, M., and Sapra, P. (2001). An expeditious solventless synthesis of isoxazoles, *Org. Prep. Proced. Int.*, **33**, pp. 381.

3. Gedye, R., Smith, F., Westaway, K., and Ali, H. (1986). The use of microwave ovens for rapid organic synthesis, *Tetrahedron Lett.*, **27**, pp. 279.

4. Rajkumar, N., Umamaheswari, D., and Ramachandran, K. (2010). Photoacoustics and magnetic studies of Fe_3O_4 nanoparticles, *Int. J. Nanoscience*, **9(3)**, pp. 243.

5. Filipponi, L., and Sutherland, D. (2010). *Nanoscience in Nature, Module 1: Fundamental Concepts in Nanoscience and Nanotechnologies* (Interdisciplinary Nanoscience Centre (iNANO), Aarhus University, Denmark).

6. Kamat, P. V. (2002). Photophysical, photochemical and photocatalytic aspects of metal nanoparticles, *J. Phys. Chem. B*, **106**, pp. 7729.

7. Liz-Marzán, L. (2006). Tailoring surface plasmons through the morphology and assembly of metal nanoparticles, *Langmuir,* **22**, pp. 32.

8. Mulvaney, P. (1996). Surface plasmon spectroscopy of nanosized metal particles, *Langmuir,* **12**, pp. 788.

9. Shipway, A. N., Katz, E., and Willner, I. (2000). Nanoparticle arrays on surfaces for electronic, optical, and sensor applications, *Chem. Phys. Chem.,* **1**, pp. 18.

10. Mallik, K., Witcomb, M. J., and Scurrell, M. S. (2005). Redox catalytic properties of gold nanoclusters: Evidence of an electron-relay effect, *Appl. Phys. A,* **80**, pp. 797.

11. Cao, Y. C., Jin, R., Nam, J., Thaxton, C. S., and Mirkin, C. A. (2003). Raman dye-labeled nanoparticle probes for proteins, *J. Am. Chem. Soc.,* **125**, pp. 14676.

12. Wang, X., Zhuang, J., Peng Q., and Li, Y. (2005). A general strategy for nanocrystal synthesis, *Nature,* **437**, pp. 121.

13. Yin, Y., and Alivisatos, P. (2005). Colloidal nanocrystal synthesis and the organic-inorganic interface, *Nature,* **437**, pp. 664.

14. Capek, I. (2004). Preparation of metal nanoparticles in water-in-oil (w/o) microemulsions, *Adv. Colloid Interface Sci.,* **110**, pp. 49.

15. Goia, D. V., and Matijevic, E. (1998). Preparation of monodispersed metal particles, *New J. Chem.,* **22**, pp. 1203.

16. Burda, C., Chen, X., Narayanan, R., and El-Sayed, M. A. (2005). Chemistry and properties of nanocrystals of different shapes, *Chem. Rev.,* **105**, pp. 1025–1102.

17. Sun, Y., and Xia, Y. (2002). Shape-controlled synthesis of gold and silver nanoparticles, *Science,* **298**, pp. 2176–2179.

18. Alagarasi. (2011). Introduction to nanomaterials. www.nccr.iitm. ac.in/2011.pdf.

19. Olson, D. H., Sheppard, E. W., McCullen, S. B., Higgins, J. B., and Scwenker, J. C. (1992). A new family of mesoporous molecular sieves prepared with liquid crystal templates, *J. Am. Chem. Soc.,* **114**, pp. 10834.

20. Ikkala, O., and Brinke, G.T. (2002). Functional materials based on self-assembly of polymeric supramolecules, *Science,* **295**, pp. 2407–2409.

21. Antonietti, M., Conrad, J., and Thunemann, A. (1994). Determination of the micelle architecture of polystyrene poly(4-vinylpyridine) block copolymers in dilute solution, *Macromolecules,* **27**, pp. 6007.

22. Vollath, D. (2013). *Nanomaterials: An Introduction to Synthesis, Properties and Application* (WILEY-VCH Verlag GmbH&Co. KGaA, Weinheim, Germany), ISBN: 978-3-527-31531-4.

23. Ma, S. K., and Lue, J. T. (1996). Spin-glass states exhibited by silver nano-particles prepared by sol-gel method, *Solid State Commun.* **97**, pp. 979.

24. Brinker, C. J., and Scherer, S. W. (1990). *Sol–Gel Science: The Physics and Chemistry of Sol–Gel Processing* (Academic Press, New York).

25. Brinker, C. J., Bunker, B. C., Tallant, D. R., Ward, K. J., and Kirkpatrick, R. J. (1988). Structure of sol–gel derived inorganic polymers: Silicates and borates, *ACS Symp. Series*, **360**, pp. 314–332.

26. Jones, R. W. (1989). *Fundamental Principles of Sol-Gel Technology* (Institute of metals, London).

27. Yang, J., Mei, S., and Ferreira, J. M. F. (2001). Hydrothermal synthesis of TiO_2 nanopowders from tetraalkylammonium hydroxide peptized sols, *Mater. Sci. Eng. C*, **15**, pp. 183–185.

28. Albrecht, M. A., Evans, C. W., and Raston, C. L. (2006). Green chemistry and the health implications of nanoparticles, *Green Chem.*, **8**, pp. 417.

29. Sun, T., and Seff, K. (1994). Silver clusters and chemistry in zeolites, *Chem. Rev.*, **94**, pp. 857.

30. Xiong, D. J., Chen, M. L., and Li, H. (2008). Colorimetric detection of pesticides based on calixarene modified silver nanoparticles in water, *Chem. Commun.*, **19(46)**, pp. 880.

31. Duran, N., Marcato, P. D., Alves, O. L., De Souza, G. I. H., and Esposito, E. (2005). Mechanistic aspects of biosynthesis of silver nanoparticles by several *Fusarium oxysporum* strains, *J. Nanobiotechnology,* **3**, pp. 8.

32. Brigger, I., Dubernet, C., and Couvreur, P. (2004). Nanoparticles in cancer therapy and diagnosis, *Adv. Drug. Deliv. Rev.*, **54**, pp. 631.

33. Guzman, M. G., Dille, J., and Godet, S. (2008). Synthesis of silver nanoparticles by chemical reduction method and their antibacterial activity, *World Acad. Sci. Eng. Technolo.*, **43**, pp. 357.

34. Zhu, Z., Kai, L., and Wang, Y. (2006). Synthesis and applications of hyperbranched polyesters: Preparation and characterization of crystalline silver nanoparticles, *Mater. Chem. Phys.*, **96**, pp. 447.

35. Sondi, I., and Salopek-Sondi, B. (2004). Silver nanoparticles as antimicrobial agent: A case study on *E. coli* as a model for Gram-negative bacteria, *J. Colloid Interface Sci.*, **275**, pp. 177.

36. Yu, D., and Yam, V. W.-W. (2004). Controlled synthesis of monodisperse silver nanocubes in water, *J. Am. Chem. Soc.*, **126**, pp. 13200.

37. Harada, M., Inada, Y., and Nomura, M. (2009). In situ time-resolved XAFS analysis of silver particle formation by photoreduction in polymer solutions, *J. Colloid Interface Sci.*, **337**, pp. 427.

38. Dubas, S. T., and Pimpan, V. (2008). Green synthesis of silver nanoparticles for ammonia sensing, *Talanta*, **76**, pp. 29.

39. Al-Thabaiti, S. A., Al-Nawaiser, F. M., Obaid, A. Y., Al-Youbi, A. O., and Khan, Z. (2008). Formation and characterization of surfactant stabilized silver nanoparticles: A kinetic study, *Colloids Surf. B Biointerfaces*, **67**, pp. 230.

40. Khan, Z., Al-Thabaiti, S. A., El-Mossalamy, E. H., and Obaid, A. Y. (2009). Preparation and characterization of silver nanoparticles by chemical reduction method, *Colloids Surf B Biointerfaces*, **73**, pp. 284.

41. Won, H., Nersisyan, H., Won, C. W., Lee, J.-M., and Hwang, J.-S. (2010). Preparation of porous silver particles using ammonium formate and its formation mechanism, *Chem. Eng. J.*, **156**, pp. 459.

42. Pastoriza, I., and Liz-marzan, L. M. (2000). Reduction of silver nanoparticles in DMF. Formation of monolayers and silver colloids, *Pure Appl. Chem.*, **72**, pp. 83.

43. Solomon, S. D., Bahadory, M., Jeyarajasingam, A. V., Rutkowsky, S. A., and Boritz, C. (2007). Synthesis and study of silver nanoparticles, *J. Chem. Edu.*, **84**, pp. 322.

44. Chang, Y. C., and Frank, C. W. (1998). Vapor deposition-polymerization of α-amino acid N-carboxy anhydride on the silicon (100) native oxide surface, *Langmuir*, **14**, pp. 326.

45. Lee, N. H., and Frank, C. W. (2003). Morphology of vapor-deposited poly (α-amino-acid) films, *Langmuir*, **19**, pp. 1295.

46. Mao, Y., and Gleason, K. K. (2004). Hot filament chemical vapor deposition of poly (glycidyl methacrylate) thin films using tert-butyl peroxide as an initiator, *Langmuir*, **20**, pp. 2484.

47. Martin, T. P., and Gleason, K. K. (2006). Combinatorial initiated CVD for polymeric thin films, *Chem. Vap. Dep.*, **12**, pp. 685.

48. Gleiter, H. (1990). Nanocrystalline materials, *Prog. Mater. Sci.*, **33**, pp. 223–315.

49. Hahn, H., and Averback, R. S. (1990). The production of nanocrystalline powders by magnetron sputtering, *J. Appl. Phys.*, **67**, pp. 1113.

50. Hahn, H., Easlman, J. A., and Siegel, R. W. (1988). *Ceramic Transactions: Ceramic Powder Science*, Vol. 1, Part B (American Ceramic Society, Westerville), pp. 1115.

51. Wu, H., and Readey, D. W. (1987). *Ceramic Transaction 2*: Silicon Carbide, 35.

52. Okabe, Y., Hojo J., and Kato, A. (1979). Formation of fine silicon carbide powders by a vapour phase method, *J. Less-Comm Metals*, 68, pp. 29.

53. Suresh, K., Prabagaran, S. R., Sengupta, S., and Shivaji, S. (2004). *Bacillus indicus* sp. nov., an arsenic-resistant bacterium isolated from an aquifer in West Bengal, India, *Int. J. Syst. Evol. Microbiol.*, **54**, pp. 1369–1375.

54. Bhainsa, K. C., and D'Souza, S. F. (2006). Extracellular biosynthesis of silver nanoparticles using the fungus *Aspergillus fumigatus*, *Colloids Surf. B Biointerfaces*, **47**, pp. 160–164.

55. Song, J. Y., and Kim, B. S. (2009). Rapid biological synthesis of silver nanoparticles using plant leaf extracts, *Bioprocess Biosyst. Eng.*, **32**, pp. 79–84.

56. Edmundson, M. C., Capeness, M., and Horsfall, L. (2014). Exploring the potential of metallic nanoparticles within synthetic biology, *N. Biotechnol.*, **31**, pp. 572–578.

57. Deplanche, K., Caldelari, I., Mikheenko, I. P., Sargent, F., and Macaskie, L. E. (2010). Involvement of hydrogenases in the formation of highly catalytic Pd(0) nanoparticles by bioreduction of Pd(II) using *Escherichia coli* mutant strains, *Microbiology*, **156**, pp. 2630–2640.

58. Yong, P., Rowson, A. N., Farr, J. P. G., Harris, I. R., and Mcaskie, L. E. (2002). Bioaccumulation of palladium by *Desulfovibrio desulfuricans*, *J. Chem. Technol. Biotechnol.*, **55**, pp. 593–601.

59. Kalimuthu, K., Suresh Babu, R., Venkataraman, D., Bilal, M., and Gurunathan, S. (2008). Biosynthesis of silver nanocrystals by *Bacillus licheniformis*, *Colloids Surf. B Biointerfaces*, **65**, pp. 150–153.

60. Johnston, C. W., Wyatt, M. A., Li, X., Ibrahim, A., Shuster, J., Southam, G., and Magarvey, N. A. (2013). Gold biomineralization by a metallophore from a gold-associated microbe, *Nat. Chem. Biol.*, **9**, pp. 241–243.

61. He, S., Guo, Z., Zhang, Y., Zhang, S., Wang, J., Gu, N. (2007). Biosynthesis of gold nanoparticles using the bacteria *Rhodopseudomonas capsulate*, *Mater. Lett.*, **61**, pp. 3984–3987.

62. Lloyd, J. R., Yong, P., and Macaskie, L. E. (1998). Enzymatic recovery of elemental palladium by using sulfate-reducing bacteria, *Appl. Environ. Microbiol.*, **64**, pp. 4607–4609.

63. Sintubin, L., De, Windt, W., Dick. J., Mast, J., van, der, Ha, D., Verstraete, W., and Boon, N. (2009). Lactic acid bacteria as reducing and capping

agent for the fast and efficient production of silver nanoparticles, *Appl. Microbiol. Biotechnol.*, **84**, pp. 741–749.

64. Sleytr, U. B., Messner, P., Pum, D., and Sára, M. (1993). Crystalline bacterial cell surface layers, *Mol. Microbiol.*, **10**, pp. 911–916.

65. Ramanathan, R., Field, M. R., O'Mullane, A. P., Smooker, P. M., Bhargava, S. K., and Bansal, V. (2013). Aqueous phase synthesis of copper nanoparticles: A link between heavy metal resistance and nanoparticle synthesis ability in bacterial systems, *Nanoscale*, **5**, pp. 2300–2306.

66. Ahmad, A., Mukherjee, P., Senapat, S., Mandal, D., Khan, M. I., Kumar, R., and Sastry, M. (2003). Extracellular biosynthesis of silver nanoparticles using the fungus *Fsarium oxysporum*, *Colloids Surf B Biointerfaces*, **28**, pp. 313–318.

67. Nair, B., and Pradeep, T. (2002). Coalescence of nanoclusters and formation of submicron crystallites assisted by *Lactobacillus* strains, *Cryst. Growth Design*, **2**, pp. 293.

68. Ingle, A., Gade, S., Pierrat, C., Sonnichsen, M. K., and Rai, M. K. (2008). Mycosynthesis of silver nanoparticles using the fungus *Fusarium acuminatum* and its activity against some human pathogenic bacteria, *Curr. Nanosci.*, **4**, pp. 141.

69. Gade, A. K., Bonde, P., Ingle, A. P., Marcato, P. D., Durán, N., and Rai, M. K. (2008). Exploitation of *Aspergillus niger* for synthesis of silver nanoparticles, *J. Biobased Mater. Bioenergy*, **2**, pp. 243.

70. Duran, N., Marcato, P. D., Alves, O., Souza, G. I., and Esposito, E. (2005). Mechanistic aspects of biosynthesis of silver nanoparticles by several *Fusarium oxysporum* strains, *J. Nanobiotechnol.*, **3**, pp. 8.

71. Kumar, S. A., Abyaneh, M. K., Gosavi, S. W., Kulkarni, S. K., Pasricha, R., Ahmad, A., and Khan, M. I. (2007). Nitrate reductase-mediated synthesis of silver nanoparticles from $AgNO_3$, *Biotechnol. Lett.*, **29**, pp. 439.

72. Kumar, S. A., Ayoobul, A. A., Absar, A., and Khan, M. I. (2007). Extracellular biosynthesis of CdSe quantum dots by the fungus, *Fusarium oxysporum*, *J. Biomedl. Nanotechnol.*, **3**, pp. 190.

Chapter 2

Synthesis of Individual Nanomaterials

Jagriti Narang,[a] **Nitesh Malhotra,**[b] **and Chandra Shekhar Pundir**[c]

[a]*Amity Institute of Nanotechnology, Amity University, Noida-201313, India*
[b]*Amity Institute of Physiotherapy, Amity University, Noida-201313, India*
[c]*Department of Biochemistry, MD University, Rohtak-124001, India*
jags_biotech@yahoo.co.in

2.1 Introduction to Noble Metal Nanomaterials

Noble metals are those metals that either do not react or react to a least extent with other elements. Nanomaterials of these noble metals, such as Au, Ag, Pt, and Pd, are exploited for various applications, for instance in the cosmetic industry, fuel cells, and analytical sensors. These have fascinating physical and chemical properties: For instance, these are biocompatible, antibacterial, and do not oxidize easily [1, 2]. Noble nanomaterials can be found in various shapes: spheres, cubes, plates, rods, wires, stars, prisms, and right bipyramids. These different shapes are exploited in various applications such as biomedicine, electronics, textile, and analytics [3–5].

Biosensors: An Introductory Textbook
Edited by Jagriti Narang and Chandra Shekhar Pundir
Copyright © 2017 Pan Stanford Publishing Pte. Ltd.
ISBN 978-981-4745-94-9 (Hardcover), 978-1-315-15652-1 (eBook)
www.panstanford.com

2.2 Parameters Affecting the Size, Dispersion, and Shape of Metal Nanoparticles

- **Reaction rate**: The size of metal nanoparticles depends on the reaction rate. The faster the reaction rate, the smaller the resulting nanoparticles.

- **Quantity of stabilizing agent**: A larger amount of stabilizing agent results in smaller nanoparticles.

- **Temperature**: High temperature results in more monodisperse particles.

- **Nature of stabilizing agent**: Weak stabilizing agents result in more monodispersed particles.

- **Differential stabilization**: Differential stabilization of crystal faces controls the shape of nanoparticles.

2.3 Synthesis of Individual Metal Nanoparticles

2.3.1 Silver Nanostructures

Silver nanoparticles can be found in various shapes, such as spheres, cubes, rods, wires, prisms, and right bipyramids. Distinct properties of silver nanoparticles make them promising nanomaterials for various applications, for instance biocompatibility, antibacterial nature, and plasmonics and surface-enhanced Raman scattering (SERS) [5]. Researchers have put efforts in synthesizing different silver nanostructures by varying the precursor concentration and stabilizer, reaction temperature, and reducing agents [6–9]. Poly vinyl pyrrolidone (PVP) acts as a stabilizer in the synthesis of silver nanomaterials. The shape of silver nanostructures can be tuned by varying the molecular weight of PVP. If it is absent or is present in less quantity, the final product is nanospheres [10]. The following methods are used for the synthesis of silver nanoparticles:

- **Chemical reduction method:** For the preparation of silver and gold nanoparticles, the chemical reduction method has been extensively explored. In this chemical approach, the precursors of respective noble metals and reducing agents are used to synthesize noble nanomaterials. Reducing agents play an important role in the synthesis of noble nanomaterials by the reduction reaction [9] (Table 2.1). By choosing different reductants, researchers can produce desired nanostructures. Various reductants are used, such as N,N-dimethylformamide (DMF), sodium borohydride ($NaBH_4$), ascorbic acid (AsA), and some reducing agents from plant source. Stabilizers promote nucleation and prevent the aggregation of nanoparticles.

- **Double reductant method:** In this approach, two reducing agents are used in different steps of preparation of nanostructures.

Table 2.1 Synthesis of different silver nanomaterials using different reducing agents

Precursor	Stabilizers	Reducing agent	Nanostructure developed
Silver nitrate ($AgNO_3$)	PVP	DMF	Nanospheres [11], nanoprisms [12, 13], and nanowires [14]
$AgNO_3$		Ethylene glycol (EG)	Cubic and right bipyramids having (100) facets [15]
$AgNO_3$	PVP/ Citrate	$NaBH_4$	Nanosized Ag colloids [16]
Double reductant method			
$AgNO_3$	PVP	EG in the first step and DMF in the second step	Silver nanoflags [17]
$AgNO_3$	PVP	$NaBH_4$ in the first step and trisodium citrate (Na_3CA) in the second step	Silver nanowires and nanomeshworks [18]

Following are the different protocols used for the preparation of silver nanostructures.

2.3.1.1 Preparation of silver nanocubes

Precursor used: $AgNO_3$

Reducing agents used: Ethylene glycol (EG) and NaBr

Silver nanocubes are prepared by the double reductant method [19]. For the synthesis of silver nanoparticles, polyvinyl pyrrolidone (PVP) is prepared in 10 ml of EG, and the mixture is heated up to 160°C for 1 h. In 1 ml of EG, 0.49 M of sodium bromide (PVP) is prepared. This prepared mixture (EG + NaBr) is added to the above solution (EG + PVP). The sonicated mixture of EG (2 ml) and $AgNO_3$ (96 mM) is added properly, and the mixture is heated up to 160°C for 1 h. The precipitates are obtained by centrifugation and washed properly [19].

2.3.1.2 Preparation of silver nanoparticles

Precursor: $AgNO_3$

Reducing agents: Sodium borohydride

Silver nanoparticles are prepared by the chemical reductant method [20]. Silver nitrate (10 ml; 1.0 mM) is added dropwise (about 1 drop/ second) to ice-cold sodium borohydride solution (30 mL; 2.0 mM), and the mixture is stirred continuously. The mixed solution develops a brighter yellow color, when all of the silver nitrate is added. Then the stirring process is stopped immediately. The solution is stable at room temperature for as long as several weeks or months.

Problems: If stirring is continued after all of the silver nitrate has been added, aggregation starts and the solution develops a darker yellow color and eventually grayish, after which the colloid starts breaking down and particles settle out.

2.3.1.3 Preparation of silver nanorods

Precursor: $AgNO_3$

Reducing agents: Sodium borohydride, ascorbic acid $(C_6H_8O_6)$

Stabilizers: Sodium citrate

The first seed solution is prepared as described by Creighton et al. [21]. First, $AgNO_3$ (2 ml; 2.5 mM) solution is prepared and citrate is mixed and NaOH (0.6 ml; 10 mM solution) is added. Then distilled water is added until a total volume of 20 ml results. Then the resulting solution is stirred continuously until a slight clouding disappears. Then the reducing agent, i.e., ice-cold $NaBH_4$ (0.6 ml of a

10 mM solution), is added and stirred for 30 s. This resulting product comprises the seed solution, which is kept for the subsequent experiment of synthesis of nanorods.

2.3.2 Gold Nanostructures

The remarkable research and advanced application of gold nanostructures have materialized only in the recent decades. Gold is a promising material for biomedical research, because of its stability and chemical inactivity; it is less prone to show any side effects in the body. Gold nanostructures possess many distinct properties such as a large surface-to-volume ratio, unique optical and electronic properties, and ease of modification. Extensive research efforts have been made to modify the surface properties of gold nanostructures for biomedical, electronic, and optical applications. The morphology, size, and stability of gold nanostructures can be tuned through different synthesis routes of nanostructures [22–24]. The gold nanostructures formed are dependent on the precursor, reducing agent, stabilizers, and also ambient conditions. The morphology of nanostructures mainly depends on two processes: nucleation and growth.

Nucleation is defined as a process involving the formation of a crystal from a solution, liquid, or vapor, in which a few ions, atoms, or molecules become arranged in a pattern characteristic of a crystalline solid, forming a site on which additional particles are deposited as the crystal grows. This process occurs spontaneously in a substance in response to a change in temperature or pressure. For nanoparticle synthesis, there should be no impurities [25].

Growth is defined as a process in which additional material deposits on this particle, causing it to increase in size [25].

2.3.2.1 Synthesis of gold nanoparticles

Precursor: Tetrachloro auric acid ($HAuCl_4$)

Reducing agent: Citric acid

Stabilizers: Citric acid

The gold in the $HAuCl_4 \cdot 3H_2O$ complex as Au^{3+} ions is reduced to neutral gold atoms, where citrate ions act as both reducing agent and a capping agent/stabilizer. In the absence of PVP or PEG, the solution becomes unstable (Table 2.2).

Table 2.2 Synthesis of different gold nanomaterials using different reducing agents

Precursors	Reducing agent	Stabilizers	Result
$HAuCl_4 \cdot 3H_2O$	Potassium bitartrate solution	Polyethylene glycol (1.0 wt%)	Measured size of nano-particles is between 50 and 200 nm [26]
$HAuCl_4 \cdot 3H_2O$	Trisodium citrate dehydrate	Trisodium citrate dehydrate	Average diameter of colloidal gold is in the range of 18 nm [27]
$HAuCl_4 \cdot 3H_2O$	Potassium bitartrate solution	PVP solution	Particle size in this experiment is less than 100 nm [26]

Gold nanoparticles are prepared according to the method of McFarland et al. [24], with a minor modification. Gold chlorauric acid solution (concentration 1 mg/5 ml; quantity 20 ml) is added. The solution is continuously stirred on a magnetic hot plate. Then the solution is brought to boiling. After that 2 ml of 1% trisodium citrate dehydrate is added dropwise. A colloidal gold solution is formed as citrate reduces gold. After some time, the color of the solution changes from blackish to wine red. This color change confirms the formation of nanoparticles. The prepared colloidal gold solution is stored in a dark bottle at 4°C.

2.3.2.2 Synthesis of gold nanorods

Gold nanorods are prepared by the seed-mediated method [28]. First, gold seeds (5 ml) are prepared. In this step, $HAuCl_4$ (250 µM) is reduced with freshly prepared ice-cold $NaBH_4$ (3 mM) in the presence of cetyl trimethylammonium bromide (CTAB; 75 mM). After mixing, the mixture develops a light brown color. Thus, gold nano-seeds are prepared, which are kept for 2 h for the subsequent preparation of nanorods. Then the growth solution (10.0 ml) is prepared by reducing $HAuCl_4$ (0.2 mM) with freshly prepared ascorbic acid (L-AA; 7 mM) in the presence of CTAB (1.6 mM). Then the solution is mixed by inversion, and the color changes from orange to colorless. Then the prepared gold seed solution is added and mixed gently. The color of the mixture becomes red gradually, which confirms the formation of gold nanorods.

2.3.2.3 Synthesis of gold nanoclusters

A nanocluster is a nanometer-sized particle made up of equal subunits.

Fluorescent gold nanoclusters are prepared by mixing a gold chloride aqueous solution (1 mM) with a biological molecule aqueous (100 mM) solution at room temperature. By varying the concentration of biological molecules, different sizes of gold nanoclusters can be produced. The mixed reaction solution is then kept at 25°C in a shaking incubator. After incubation, a dark red solution is obtained. The prepared mixture contains both gold nanoparticles and fluorescent gold nanoclusters. The gold nanoparticles are removed from the mixture by centrifugation (15,000 rpm), resulting in a clear gold nanocluster solution [29].

2.3.3 Synthesis of Iron Oxide/Magnetic Nanostructures

The synthesis of iron oxide magnetic nanoparticles involves three methods: microemulsion, thermal decomposition, and coprecipitation. Among them, the chemical coprecipitation method is convenient, economical, and high yielding. This method offers a low-temperature alternative to the conventional powder synthesis techniques in the production of nanoparticles, and the sizes of nanoparticles can be well controlled by an apt surfactant [30]. Chemical coprecipitation can produce fine, high-purity stoichiometric particles of single and multicomponent metal oxides [31]. Parameters that can be varied to produce the desired nanoparticles are solution pH, temperature, stirring rate, solute concentration, and surfactant concentration. The surfactant helps in preventing the agglomeration of the iron nanoparticles upon exposure to air. The chemical method involves the coprecipitation of iron salts, i.e., Fe(II) and Fe(III) ions, by aqueous ammonia or sodium hydroxide [32, 33]. Iron nanoparticles can also be produced by the thermal decomposition of an alkaline solution of Fe (III) chelate in the presence of hydrazine [34]. The principal advantage of this method is that a large amount of nanoparticles can be synthesized, but tuning of the desired particle size is restricted. It is due to the fact that only kinetic factors affect the size of nanoparticles. The size and shape of the nanoparticles can be tailored by adjusting kinetic factors such as pH, concentration, temperature, nature of the precursor salts, or the Fe(II)/Fe(III) concentration ratio. The higher the pH and ionic strength, the smaller the particle size and size distribution width, because these parameters determine the chemical composition of the crystal surface and consequently the electrostatic surface charge of the particles [35].

2.3.3.1   Synthesis of iron oxide nanoparticles

Precursor: Ferrous chloride tetrahydrate $(FeCl_2 \cdot 4H_2O)$, ferric chloride hexahydrate

Surfactant: Sodium hydroxide (NaOH)

Iron oxide nanoparticles are prepared according to the method of Predoi [36]. In this method, 0.5 M ferrous chloride tetrahydrate $(FeCl_2 \cdot 4H_2O)$ is prepared in 2 M HCl, and 0.5 M ferric chloride hexahydrate $(FeCl_3 \cdot 6H_2O)$ is prepared in DW and the two are mixed at room temperature. This prepared mixture is added to NaOH (200 ml; 1.5 M) solution with continuous stirring for about half an hour. The resulting precipitates are obtained by centrifugation at 8000×g and dried at 40°C.

2.3.3.2   Synthesis of magnetic nanorods

Precursors: Ferrous ammonium sulfate $[(NH_4)_2Fe(SO_4)_2]$

Surfactant: Polyethylene glycol (PEG)

Ferrous ammonium sulfate (2.5 g) is added to distilled water (60 mL) saturated with oxygen. Then PEG 2000 (6 mL) is added to the resulting mixture. Aqueous ammonia is placed in the right 100 ml flask. Ammonia is isothermally diffused to the left flask after opening the stopcock. The solution is not shaken and stirred during the diffusion, and both flasks are heated at certain temperatures. Then the solution is cooled to room temperature. The resulting precipitates are collected and washed with absolute ethanol and DW to remove PEG. The final products are dried at 333 K [37].

During the preparation, two main reactions take place in the solution: deposition and oxidation of Fe(II) ions. At the pH condition of Fe(OH)$_2$ formation, the Fe(III) must deposit as soon as it is formed from the oxidation of Fe(II) because of the great difference between the Ksps of Fe(OH)$_2$ (pKsp = 15.1) and Fe(OH)$_3$ (pKsp = 37.4) [37].

Another route for the preparation of magnetic nanoparticles is the colloidal dispersion of inorganic and organic hybrid materials. This method has several advantages over the chemical coprecipitation method, such as good control over shape and size, homogeneous multicomponent systems, and requirement of low temperature. In this method, two processes are involved: hydroxylation and condensation.

2.3.3.3 Synthesis of iron oxide nanoparticles through sol–gel techniques

First, the gel is prepared by condensing ferric chloride in the presence of an alkali such as NaOH or KOH. Then this condensed ferric hydroxide is heated at 100°C for 8 days and monodispersed pseudo-cubic α-Fe$_2$O$_3$ particles are obtained [38].

Another approach for the synthesis of iron oxide nanoparticles is the hydrothermal technique, which is defined as any heterogeneous reaction in the presence of mineralizers under high pressure and temperature conditions. In this case, a heterogeneous reaction occurs and high temperature is required during processing.

2.3.3.4 Synthesis of iron oxide nanoparticles through hydrothermal techniques

First, iron chloride solution and iron powder in the presence of urea are treated hydrothermally at 130–150°C for 20 h. The resulting product is rod-shaped magnetite particles of about 80 nm [39].

2.3.4 Synthesis of Zinc Oxide Nanomaterials

Zinc oxide exhibits distinct physical and chemical properties such as high surface area, high catalytic efficiency, non-toxicity, chemical stability, and strong adsorption ability (high isoelectric point, pH 9.5), high chemical stability, high electrochemical coupling coefficient, broad range of radiation absorption, and high photostability [40–42]. Nanoporous ZnO nanoparticles have low toxicity. Biocompatibility and a large active surface area for strong adsorption of biomolecules make these a promising material for biological applications [42]. Different methods are available for the synthesis of ZnO nanomaterials because of their vast applications: chemical precipitation, vapor deposition, hydrothermal synthesis, sol–gel process, and precipitation from microemulsions.

2.3.4.1 Synthesis of zinc oxide nanorods by sol–gel technique

First, zinc acetate dihydrate solution is prepared by dissolving 2.195 g of zinc acetate dihydrate in 100 ml distilled water/ethanol, and stirred in ambient atmosphere conditions. Then potassium hydroxide (1.122 g dissolved in 10 ml distilled water) is added to the just prepared solution dropwise under continuous stirring. After a few minutes, a milky white solution is obtained. This prepared mixture is further heated for 3 h at 80–90°C without stirring. The resulting suspension is centrifuged to retrieve the product, and the mixture is washed with distilled water in an ultrasonic bathwater and then the powder is dried at 70°C overnight [43].

2.3.4.2 Synthesis of zinc oxide nanoflower

Zinc oxide nanoflower is prepared through the hydrothermal technique. Zinc nitrate (0.025 M) and hexamethyltetramine solution is prepared at a molar ratio of 1:1. Afterward, the substrates or support are dipped into the mixture. Subsequently, the solution with the dipped substrate is heated at a temperature of 90°C for 1 h with continuous stirring. After deposition, the substrate with deposited nanostructures is washed with distilled water and dried in air at room temperature [43].

The control of the as-fabricated ZnO flowerlike structures can be achieved by altering hydrothermal growth conditions.

2.3.4.3 Synthesis of zinc oxide nanoparticles

Precursors: Zinc nitrate, zinc acetate

Zinc oxide nanoparticles are prepared by adding NaOH (0.9 M) to zinc nitrate solution (0.45 M) dropwise. The solution is heated at

55°C with continuous stirring. Afterward the resulting solution is probe sonicated at higher amplitude for 30 min and dried in vacuum at 60°C [44].

2.3.4.4 Synthesis of zinc oxide nanobelts

Precursor: $ZnCl_2$

Zinc oxide nanobelts are prepared by dissolving zinc chloride (0.2 g), sodium dodecyl sulfonate (1.5 g), and sodium bicarbonate (20 g) to make the final solution (4.72 M). The prepared mixture is subsequently added to a Teflon-lined stainless steel autoclave. The resulting reaction mixture is continuously stirred for half an hour. The autoclave is sealed and maintained at 140°C for 12 h. Following the autoclave, the resulting precipitated white powder is filtered and washed with ethanol. Then the resulted mixture is dried in vacuum at 60°C for 4 h [45].

2.3.4.5 Synthesis of zinc oxide nanoclusters

Precursor: Zinc acetate

Zinc oxide nanoclusters are prepared by dissolving zinc acetate (2.195 g) in 0.1 M ethanol and are refluxed for 3 h at 80°C. Lithium hydroxide (0.586 g) is dissolved in 100 ml distilled water. After the reflux of zinc acetate, the condensate is separated, and the prepared lithium hydroxide is added dropwise with continuous stirring. The precipitate is separated by centrifugation at 7000 rpm for 15 min, and then the samples are dried at 80°C for 12 h [46].

2.3.5 Synthesis of Carbon-Based Nanomaterials

Carbon-based nanomaterials are presently one of the most striking nanomaterials, as these are exploited for many applications [47]. Carbon nanomaterials include single- and multi-walled carbon nanotubes (CNTs), carbon nanoparticles, and graphene nanoparticles. Carbon nanomaterials exhibit electronic and mechanical properties, which are exploited in various applications such as in optics [48], electronics [49], and biomedicine [47]. Various methods are available for synthesizing carbon-based nanomaterials, such as laser ablation [50] or arc discharge method [51]. In these approaches, the synthesized carbon nanomaterials have to be purified, and the purified nanomaterials are suspended in an organic solvent for deposition onto a desired substrate. The disadvantage of this method is that controlling the orientation of nanotubes on the solid substrate becomes complicated. One of the approaches that can satisfy the requirements of cost, low temperature, orientation control, and high purity is chemical vapor deposition (CVD).

The plasma-enhanced CVD (PECVD) method can be used to synthesize carbon nanomaterials for commercialization. The advantage of this method is that the nanomaterials are easily deposited on the substrate and are easy to collect [52]. Plasmas used to activate the reactants in the gas phase are usually generated by hot filaments or by electrical discharges at different frequencies. In PECVD synthesis, the catalyst powder is generally loaded on a substrate by means of wet chemistry, generally accompanied by chemical etching to induce the formation of catalyst particles on the substrate.

2.3.5.1 Synthesis of carbon nanotubes

Synthesis methods of CNTs can be chemical and physical. The chemical method removes impurities and purifies CNTs, since impurities present are oxidized at a faster rate than CNTs. Chemical oxidation can be through either gas, such as O_2 and Cl_2, or liquid using acid treatment and refluxing. The physical method involves the separation of CNTs on the basis of differences in their physical size, aspect ratio, gravity, and magnetic properties.

The chemical method involves three processes for the synthesis of CNTs for both tip and base growth. First, a round precursor is formed on the surface of the substrate. The second process involves the diffusion of carbon on the sides of the precursor, but the precursor remains at the base and causes hollow core of the nanotube. The third process involves the formation of rod-like carbon structure. In this process, the nanotube grows in upward direction, while metal remains attached to the substrate. In the tip growth method, the particle detaches itself from the substrate and resides on the top of the growing nanotube. Depending on the size of the catalyst particles, single-walled or multi-walled nanotubes are grown [53].

For the synthesis of CNTs, the substrate or membrane, with or without the Ni catalyst in the pores, is placed on a right-angled platinum grid in the CVD reactor. In the case of without Ni, temperature is increased to 900°C, while with the Ni catalyst, temperature is increased to 545°C, both under an argon flow. The carbon nanofibers and nanotubes are synthesized by the decomposition of ethylene. With ethylene, the argon flow is terminated after the temperature stabilizes. Simultaneously, a 10 sccm flow of ethylene is initiated. After deposition, the argon flow is resumed, the ethylene flow is terminated, and the furnace is turned off and allowed to cool to room temperature [54].

For the formation of single-walled CNTs, researchers have used SiO_2, MgO as support and Ni, Fe-Co, respectively, as catalyst [55, 56]. For the formation of multi-walled nanotubes, researchers have used Al_2O_3 and $CaCO_3$ as support and Fe-Co, Ni as catalyst [57, 58].

2.3.5.2 Preparation of graphene oxide nanoparticles

Graphene oxide is chemically exfoliated from oxidized graphite. It has some exceptional properties due to its chemical structure, which is composed of sp^3 carbon domains surrounding sp^2 carbon domain, and a wide range of functional groups such as epoxy, hydroxyl, and carboxyl [59–61].

Graphene oxide nanoparticles are prepared according to a modified Hummers–Offeman method, in which graphite powder is used as the precursor and sulfuric acid acts as the oxidizing agent.

Graphite powder (0.5 g) is dissolved in sulfuric acid (23 ml) at 4°C. Then sodium nitrate ($NaNO_3$; 0.5 g) is added followed by 10 ml of potassium permanganate ($KMnO_4$; 2 mM) dropwise. The mixture is stirred continuously for 1 h in a 35°C water bath. Then, H_2O (140 ml) is added to the mixture, and the temperature is raised further to 90°C. Afterward, H_2O_2 (3 ml; 30 wt%) is injected into the mixture, and the color of the mixture changes to light brown. The solution is then centrifuged at 4000 rpm to obtain graphene oxide nanoparticles. The precipitates are then filtered and washed.

2.3.5.3 Preparation of graphene nanoflakes

Graphene nanoflakes are prepared after some modification in the protocol of literature [21]. First, sodium dodecyl benzene sulfonate (SDBS) surfactant (20 mg/ml) is prepared in distilled water with continuous stirring for 12 h. Then graphite powder (10 mg) and SDBS (15 ml) are mixed together and then subjected to ultrasonication for 30 min. The resulting solution is centrifuged at 1000 rpm for 45 min. After centrifugation, the supernatant is collected and sonicated for 20 min. Thus, graphene nanoflakes are obtained.

2.3.5.4 Preparation of carbon nanospheres

Kerosene oil is poured in a laboratory lamp, and the lamp is lighted and allowed to burn in open air. The electrode/substrate is placed above the flame of the lamp in order to collect the soot emitted from the lamp. Then the electrode is allowed to dry. The electrode with a deposition of carbon nanospheres is obtained (CNS/FTO).

2.4 Conclusion

The huge demand for nanomaterials of various shapes in different fields of sciences as bioalternatives, which can perform better than bulk materials, has fascinated the interest of researchers for synthesizing different morphologically advanced novel nanomaterials. Based on this, there are descriptions of synthesis methods of various nanomaterials with different morphologies. Apart from this, the essential necessities of nanomaterials have also been discussed. Comprehensive information has been enclosed regarding the precursors and reducing agents required for the synthesis of nanomaterials with different shapes.

Problems

1. What is the major problem associated with Ag nanoparticle synthesis?
2. Explain the difference between a nanowire and a nanorod.
3. Explain various techniques involved in the synthesis of nanomaterials.
4. Explain the parameters that affect the size of nanoparticles?
5. Explain nucleation and growth?

References

1. Butun, S., and Sahiner, N. (2011). A versatile hydrogel template for metal nanoparticle preparation and their use in catalysis, *Polymer*, **52**, pp. 4834–4840.

2. Harish, S., Sabarinathan, R., Joseph, J., and Phani, K.L.N. (2011). Role of pH in the synthesis of 3-aminopropyl trimethoxysilane stabilized colloidal gold/silver and their alloy sols and their application to catalysis, *Mater. Chem. Phys.*, **127**, pp. 203–207.

3. Cao, Y., Li, D., Jiang, F., Yang, Y., and Huang, Z. (2013). Engineering metal nanostructure for SERS application, *J. Nanomater.*, pp. 1–12.

4. Botta, R., Upender, G., Sathyavathi, R., Narayana Rao, D., and Bansal, C. (2013). Silver nanoclusters films for single molecule detection using Surface Enhanced Raman Scattering (SERS), *Mater. Chem. Phys.*, **137**, pp. 699–703.

5. Zhu, S.Q., Zhang, T., Guo, X.L., Wang, Q.L., Liu, X.F., and Zhang, X.Y. (2012). Gold nanoparticles thin films fabricated by electrophoretic deposition method for highly sensitive SERS application, *Nanoscale Res. Lett.*, **7**, pp. 1–7.

6. Wiley, B., Sun, Y.G., Chen, J.Y., Cang, H., Li, Z.Y., Li, X.D., and Xia, Y.N. (2005). Shape-controlled synthesis of silver and gold nanostructures, *Mrs Bull.*, **30**, pp. 356–361.

7. Sun, Y., and Xia, Y. (2002). Shape-controlled synthesis of gold and silver nanoparticles, *Science*, **298**, pp. 2176–2179.

8. Wiley, B., Sun, Y.G., Mayers, B., and Xia, Y.N. (2005). Shape-controlled synthesis of metal nanostructures: The case of silver, *Chem.-A Eur. J.*, **11**, pp. 454–463.

9. Xia, Y., Xiong, Y., Lim, B., and Skrabalak, S.E. (2009). Shape-controlled synthesis of metal nanocrystals: Simple chemistry meets complex physics? *Angew. Chem. Int. Ed. Engl.*, **48**, pp. 60–103.

10. Song, Y.J., Wang, M.L., Zhang, X.Y., Wu, J.Y., and Zhang, T. (2014). Investigation on the role of the molecular weight of polyvinyl pyrrolidone in the shape control of high-yield silver nanospheres and nanowires, *Nanoscale Res. Lett.*, **9**, pp. 1–8.

11. Pastoriza-Santos, I., and Liz-Marzán, L.M. (2002). Formation of PVP-protected metal nanoparticles in DMF, *Langmuir*, **18**, pp. 2888–2894.

12. Pastoriza-Santos, I., and Liz-Marzán, L.M. (2002). Synthesis of silver nanoprisms in DMF, *Nano Lett.*, **2**, pp. 903–905.

13. Pastoriza-Santos, I., and Liz-Marzán, L.M. (2009). *N,N*-Dimethylformamide as a reaction medium for metal nanoparticles synthesis, *Adv. Funct. Mater.*, **19**, pp. 679–688.

14. Giersig, M., Pastoriza-Santos, I., and Liz-Marzán, L.M. (2004). Evidence of an aggregative mechanism during the formation of silver nanowires in *N,N*-dimethylformamide, *J. Mater. Chem.*, **14**, pp. 607–610.

15. Chang, S., Chen, K., Hua, Q., Ma, Y., and Huang, W. (2011). Evidence for the growth mechanisms of silver nanocubes and Nanowires, *J. Phys. Chem. C*, **115**, pp. 7979–7986.

16. Zhang, Y., Yang, P., and Zhang, L. (2012). Size- and shape-tunable silver nanoparticles created through facile aqueous synthesis, *J. Nanopart. Res.*, **15**, pp. 1-10.

17. Tsuji, M., Tang, X., Matsunaga, M., Maeda, Y., and Watanabe, M. (2010). Shape evolution of flag types of silver nanostructures from nanorod seeds in PVP-assisted DMF solution, *Cryst. Growth Des.*, **10**, pp. 5238–5243.

18. Zhang, X.Y., Zhang, T., Zhu, S.Q., Wang, L.D., Liu, X.F., Wang, Q.L., and Song, Y.J. (2012). Fabrication and spectroscopic investigation of branched silver nanowires and nanomeshworks, *Nanoscale Res. Lett.*, **7**, pp. 596.

19. Andrew, R.S., Joseph, M.M., Jingyi, C., and Younan, X. (2006). Rapid synthesis of small silver nanocubes by mediating polyol reduction with a trace amount of sodium sulfide or sodium hydrosulfide, *Chem. Phys. Lett.*, **432**, pp. 491–496.

20. Solomon, S.D., Bahadory, M., Jeyarajasingam, A.V., Rutkowsky, S.A., and Boritz, C. (2007). Synthesis and study of silver nanoparticles, *J. Chem. Educ.*, **84(2)**, pp. 322.

21. Creighton, J.A., Blatchford, C.G., and Grant-Albrecht, M. (1979). Plasma resonance enhancement of Raman scattering by pyridine adsorbed on silver or gold sol particles of size comparable to the excitation wavelength, *J. Chem. Soc. Faraday Trans.*, **75**, pp. 790.

22. Daniel, M.-C., and Astruc, D. (2003). Gold nanoparticles: Assembly, supramolecular chemistry, quantum-size-related properties, and applications toward biology, catalysis, and nanotechnology, *Chem. Rev.*, **104**, pp. 293–346.

23. Faraday, M. (1857). The Bakerian Lecture: Experimental relations of gold (and other metals) to light, *Philos. Trans. R. Soc. Lond.*, **147**, pp. 145–181.

24. Grzelczak, M., Perez-Juste, J., Mulvaney, P., and Liz-Marzan, L.M. (2008). Shape control in gold nanoparticles synthesis, *Chem. Soc. Rev.*, **37**, pp. 1783–1791.

25. Turkevich, H. (1949). Electron microscopy of colloidal systems, *Anal. Chem.*, **21**, pp. 475.

26. Khalida, S.M., Hadi, D.A., Zaid, M.A., and Hashim, A.Y. (2012). Comparative study on methods for preparation of gold nanoparticles, *Green Sustainable Chem.*, **2**, pp. 26–28.

27. Verma, H.N., Singh, P., and Chavan, R.M. (2014). Gold nanoparticles: Synthesis and characterization, *Veterinary World*, **7(2)**, pp. 72–77.

28. Chen, C.-D., Cheng, S.-F., Chau, L.-K., Wang, C.R.C. (2007). Sensing capabilities of the localized surface plasmon resonance of gold nanorods. *Biosens. Bioelectron.*, **22**, pp. 926–932.

29. Wintzinger, L., An, W., Turner, C.H., and Bao, Y. (2010). Synthesis and modeling of fluorescent gold nanoclusters, *Fluorescent Gold Nanoclusters*, **7**, pp. 24–27.

30. Nidhin, M., Indumathy, R., Sreeram, K.J., and Nair, B.U. (2008). Synthesis of iron oxide nanoparticles of narrow size distribution on polysaccharide templates, *Bull. Mater. Sci.*, **31**, pp. 93–96.

31. Chen, S., Feng, J., Guo, X., Hong, J., and Ding, W. (2005). One-step wet chemistry for preparation of magnetite nanorods, *Mater. Lett.*, **59**, pp. 985–988.

32. Kang, Y.S., Risbud, S., Rabolt, J.F., and Stroeve, P. (1996). Synthesis and characterization of nanometer-size Fe_3O_4 and g-Fe_2O_3 particles, *Chem. Mater.*, **8**, pp. 2209.

33. Hong, C.Y., Jang, I.J., Horng, H.E., Hsu, C.J., Yao, Y.D., and Yang, H.C. (1997). Ordered structures in FeO kerosene-based ferrofluids, *J. Appl. Phys.*, **81**, pp. 4275.

34. Vijayakumar, R., Koltypin, Y., Felner, I., and Gedanken, A. (2000). Sonochemical synthesis and characterization of pure nanometer-sized Fe_3O_4 particles, *Mater. Sci. Eng.*, **286**, pp. 101.

35. Tartaj, P., Morales, M.P., Veintemillas-Verdaguer, S., Gonzalez-Carreno, T., and Serna, C.J. (2006). Synthesis, properties and biomedical applications of magnetic nanoparticles. *Handbook of Magnetic Materials* (Elsevier, Amsterdam, the Netherlands), pp. 403.

36. Predoi, D. (2007). A study on iron oxide nanoparticles coated with dextrin obtained by coprecipitation, *Dig. J. Nanomater. Bios.*, **2**, pp. 169–173.

37. Lian, S., Wang, E., Kang, Z., Bai, Y., Gao, L., Jiang, M., Hu, C., and Xu, L. (2004). Synthesis of magnetite nanorods and porous hematite nanorods, *Solid State Commun.*, **129**, pp. 490.

38. Sugimoto, T., and Sakata, K. (1992). Preparation of monodisperse pseudocubic α-Fe$_2$O$_3$ particles from condensed ferric hydroxide gel, *J. Colloid Interf. Sci.*, **152**, pp. 587–590.

39. Yitai, Q., Yi, X., Chuan, H., Jing, L., and Zuyao, C. (1994). Hydrothermal preparation and characterization of ultrafine magnetite powders, *Mat. Res. Bull.*, **29**, pp. 953–957.

40. Segets, D., Gradl, J., Taylor, R.K., Vassilev, V., and Peukert, W. (2009). Analysis of optical absorbance spectra for the determination of ZnO nanoparticle size distribution, solubility, and surface energy, *ACS Nano*, **3**, pp. 1703–1710.

41. Lou, X. (1991). Development of ZnO series ceramic semiconductor gas sensors, *J. Sens. Trans. Technol.*, **3**, pp. 1–5.

42. Narang, J., and Pundir, C.S. (2011). Construction of a triglyceride amperometric biosensor based on chitosan–ZnO nanocomposite film, *Int. J. Biol. Macromol.*, **49**, pp. 707–715.

43. Bhakat, C., and Singh, P.P. (2012). Zinc oxide nanorods: Synthesis and its applications in solar cell, *Intl. J. Modern Engg. Res.*, **2(4)**, pp. 2452–2454.

44. Samanta, P.K., Patra, S.K., Ghosh, A., and Chaudhuri, P.R. (2009). Visible emission from ZnO nanorods synthesized by a simple wet chemical method, *Intl. J. NanoSc. Nanotech.*, **1(1-2)**, pp. 81–90.

45. Hanmei, H., Xianhuai, H., Chonghai, D., Xiangying, C., and Yitai, Q. (2007). Hydrothermal synthesis of ZnO nanowires and nanobelts on a large scale, *Mater. Chem. Phys.*, **106**, pp. 58–62.

46. Kaur, J., Kumar, P., Thangaiah, S.S., and Thangaraj, R. (2013). Structural, optical and fluorescence properties of wet chemically synthesized ZnO: Pd^{2+} nanocrystals, *Int. Nano Lett.*, **3**, pp. 4.

47. Huczko, A. (2002). Synthesis of aligned carbon nanotubes, *Appl. Phys. Mater. Sci. Process.*, **74(5)**, pp. 617–638.

48. Miyagawa, H., Misra, M., and Mohanty, A.K. (2005). Mechanical properties of carbon nanotubes and their polymer composites, *J. Nanosci. Nanotechnol.*, **5(10)**, pp. 1593–1615.

49. Jeong, W., and Kessler, M.R. (2008). Toughness enhancement in romp functionalized carbon nanotube/polydicyclopentadiene composites, *Chem. Mater.*, **20(22)**, pp. 7060–7068.

50. Thess, A., Lee, R., Nikolaev, P., Dai, H.J., Petit, P., Robert, J., Xu, C.H., Lee, Y.H., Kim, S.G., Rinzler, A.G., Colbert, D.T., Scuseria, G.E., Tomanek, D.,

Fischer, J.E., and Smalley, R.E. (1996). Crystalline ropes of metallic carbon nanotubes, *Science,* **273**, pp. 483–487.

51. Journet, C., Maser, W.K., Bernier, P., Loiseau, A., Lamy de la Chapelle, M., Lefrant, S., Deniard, P., Lee, P., and Fischer, J.E. (1997). Large-scale production of single-walled carbon nanotubes by the electric-arc technique, *Nature,* **388**, pp. 756–758.

52. Duesberg, G.S., Graham, A.P., Kreupl, F., Liebau, M., Seidel, R., Unger, E., Hoenlein, W. (2004). Ways towards the scaleable integration of carbon nanotubes into silicon based technology, *Diamond Relat. Mater.*, **13**, pp. 354–361.

53. Antisari, M.A., Marazzi, R., and Krsmanovic, R. (2003). Synthesis of multiwall carbon nanotubes by electric arc discharge in liquid environments, *Carbon,* **41**, pp. 2393–2401.

54. Che, G., Lakshmi, B.B., Martin, C.R., and Fisher, E.R. (1998). Chemical vapor deposition based synthesis of carbon nanotubes and nanofibers using a template method department of chemistry, *Chem. Mater.*, **10**, pp. 260–267.

55. Qi., X., Qin., C., Zhong, W., Au., C., Ye., X., and Du, Y. (2010). Large-scale synthesis of carbon nanomaterials by catalytic chemical vapor deposition: A review of the effects of synthesis parameters and magnetic properties, *Materials,* **3**, pp. 4142–4174.

56. Li, C.Y., Zhu, H.W., Suenaga, K., Wei, J.Q., Wang, K.L., and Wu, D.H. (2009). Diameter dependent growth mode of carbon nanotubes on nanoporous SiO_2 substrates, *Mater. Lett.,* **63**, pp. 1366–1369.

57. Lee, W.Y., Lin, H., Gu, L., Leou, K.C., and Tsai, C.H. (2008). CVD catalytic growth of single-walled carbon nanotubes with a selective diameter distribution, *Diamond Relat. Mater.,* **17**, pp. 66–71.

58. Xu, Y., Li, Z.R., Dervishi, E., Saini, V., Cui, J.B., Biris, A.R., Lupu, D., and Biris, A.S. (2008). Surface area and thermal stability effect of the MgO supported catalysts for the synthesis of carbon nanotubes, *J. Mater. Chem.,* **18**, pp. 5738–5745.

59. Loh, K.P., Bao, Q., Eda, G., and Chhowalla, M. (2010). Graphene oxide as a chemically tunable platform for optical applications, *Nat. Chem.,* **2**, pp. 1015–1024.

60. Nakada, K., Fujita, M., and Dresselhaus, G. (1996). Edge state in graphene ribbons: Nanometer size effect and edge shape dependence, *Phys. Rev. B,* **54**, pp. 17954–17961.

61. He, H., Klinowski, J., and Forster, M. (1998). A new structural model for graphite oxide, *Chem. Phys. Lett.,* **287**, pp. 53–56.

Chapter 3

Characterization Techniques

Jagriti Narang,[a] Nitesh Malhotra,[b] and Rachna Rawal[a]

[a]*Amity Institute of Nanotechnology, Amity University, Noida-201313, India*
[b]*Amity Institute of Physiotherapy, Amity University, Noida-201313, India*
jags_biotech@yahoo.co.in

3.1 Introduction

For the characterization of nanomaterials, high magnification and resolution are required, which cannot be achieved through optical microscopes. Thus, nanomaterial characterization requires more resolution, which can be achieved by transmission electron microscopy (TEM), scanning electron microscopy (SEM), scanning tunneling microscopy (STM), and atomic force microscopy (AFM). These imaging techniques help in observing particles in the range of even picometer with high resolution. All these imaging techniques facilitate in producing magnified and clear images. SEM and TEM illustrate two-dimensional view of nanomaterials, while AFM depicts three-dimensional views.

For the characterization of nanomaterials, researchers exploited these imaging techniques to prove their results. These techniques

Biosensors: An Introductory Textbook
Edited by Jagriti Narang and Chandra Shekhar Pundir
Copyright © 2017 Pan Stanford Publishing Pte. Ltd.
ISBN 978-981-4745-94-9 (Hardcover), 978-1-315-15652-1 (eBook)
www.panstanford.com

are extensively useful for visualizing nanoparticles [1]. The basic theory for all these imaging techniques is same, but principles are different.

3.2 Scanning Electron Microscopy

Electron microscopes use a beam of energetic electrons to scan substances on a very fine scale. SEM can provide shape, morphology, particle number, and size of nanomaterials. It portrays nanomaterials in two dimensions with high magnification and resolution. It also gives information about surface topography and composition of nanomaterials if an energy dispersive detector (EDX) is attached [2]. The composition of nanoparticles can be easily estimated using SEM in conjunction with EDX. If the nanomaterial is composed of heavy metals such as Au and Ag, it can be easily detected through this technique, but low–molecular weight component cannot be identified using EDX technique [3].

3.2.1 Principle

In SEM, an electron beam is focused on the specimen surface. In this instrument, when the electron beam is incident on the specimen surface, it generates signals in the form of electrons or photons. Signals emitted from the specimen are collected by detectors and are displayed on a screen. The energy dispersive X-ray detector can sort the X-ray signals by energy and produce elemental images, so the spatial distribution of particular elements can be detected by SEM (Fig. 3.1).

3.2.2 Parameter Study through SEM

Morphological and sizing analysis, insight on the degree of aggregation of nanoparticles, position of nanoparticles over the substrate, and high degree of dispersion and uniformity of metallic nanoparticles over the substrate can be studied through SEM.

3.2.3 Sample Preparation

Nanoparticles should be dry and should be conductive for the electron beam. This dry sample is then mounted on a sample holder.

If the nanoparticles are nonconductive, they are coated with a conductive metal, such as gold, by sputter coating. It is done to avoid charge effect on the specimen during electron irradiation. In the case of polymer nanocomposites, the polymer should be able to hold up the condition of vacuum; otherwise, the polymer can be damaged.

3.2.4 Disadvantages

This process is time consuming and costly, and it requires desired ambient conditions and skilled persons to operate the instrument.

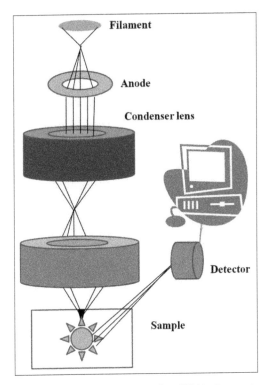

Figure 3.1 Diagrammatic representation of an SEM instrument.

3.3 Transmission Electron Microscopy

TEM provides information about particle size, shape, and any surface layers [4, 5]. Any change in the morphology or shape of nanoparticles can be identified through this technique [6]. TEM provides highly magnified images with resolutions close to 0.1 nm. Consequently, the study of nanoparticles can be significantly enhanced by this technique [7, 8]. With better resolution, a new level of understanding of the behavior of matter at the nanoscale can be achieved. High-resolution transmission electron microscopy (HRTEM) is another imaging mode of the transmission electron microscope that allows the imaging of the crystallographic structure of a sample at the atomic scale.

3.3.1   Principle

In TEM imaging, the specimen or nanoparticles interact with the electron beam mostly by diffraction, rather than absorption. Before interaction, the electron beam is restricted by the condenser aperture to remove high-angle electrons. After interaction, detection signals are produced through scattered electrons, along with some unscattered electrons. The signals generated contain all information about the sample (Fig. 3.2).

3.3.2   Parameter Study through TEM

Particle size, shape, crystallinity, and interparticle interaction [8, 9] can be studied using TEM. Through HRTEM, we can also detect individual atoms and crystalline defects.

3.3.3   Disadvantages

Image overlapping is a disadvantage of this technique (surrounding matrix tends to mask the nanoparticles). Nanoparticles can be susceptible to damage under the high electron beam irradiation.

3.3.4   Sample Preparation

Nanoparticles should be dry since water can collapse the sample under vacuum and should be conductive for the electron beam. The sample should be extremely thin so that the electron beam can easily pass through the sample.

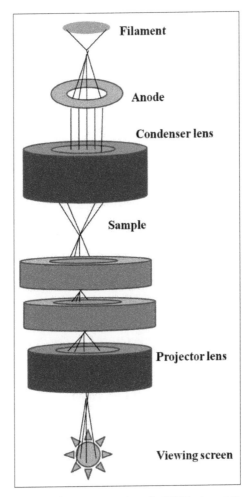

Figure 3.2 Diagrammatic representation of a TEM instrument

3.4 Atomic Force Microscopy

In scanning probe microscopies, AFM is included [10]. This technique overcomes some limitations of SEM and TEM. It can be applied to nonconductive nanoparticles; no dehydration is required; and ambient environmental conditions are sufficient [11]. AFM provides a three-dimensional view of the nanomaterial. Moreover, this techniques allows the investigation of particles in their own environment [10]. There are different approaches to AFM: (i) contact, (ii) non-contact, and (iii) tapping scanning [12]. The contact mode is the most usual imaging approach where the cantilever is deflected as it moves over the surface. The tip is constantly adjusted to maintain a constant deflection to "read" the sample topography. Due to the dragging motion of the tip, nanoparticles that are weakly attached to the substrate surface can be damaged [13]. To minimize the defects caused by the contact approach, non-contact and tapping modes are employed as in these approaches, only the tip touches the surface momentarily, minimizing the physical contact. This technique has the advantage of imaging any type of surface, including polymers, ceramics, composites, glass, and biological samples. Furthermore, this technique has one more benefit: the voltage and the tip-to-substrate spacing can be tuned independently.

3.4.1 Principle

In a cantilever, a sharp probe tip scans over the nanoparticles [14]. The probe tip shifts up and down in reaction to forces of attraction or repulsion with the specimen surface. The movement of the cantilever is detected by a laser and photodetector.

3.4.2 Parameter Study through AFM

Extraordinary topographic features such as volume, height, size, shape, aspect ratio, and particle surface morphology can be studied with AFM. Both SEM and AFM techniques are employed for the investigation of surface topography, but SEM is best for chemical composition while AFM is best for mechanical properties of a sample.

3.4.3 Disadvantages

This technique is time consuming, and the specimen should be fixed over the substrate to avoid distortion due to movement.

3.4.4 Sample Preparation

For rough surfaces, the contact approach of AFM is preferred, while for soft samples, the tapping or non-contact approach of AFM is preferred.

3.5   Scattering Methods

In scattering methods, a radiation can be either light or sound, which turns aside from a straight path because of the specimen or nanoparticles suspended in the medium through which the radiation wave travels.

3.5.1   Light Scattering

There are two types of light scattering methods: (i) dynamic and (ii) static. Static light scattering measures time–average intensities, while dynamic light scattering (DLS) measures real-time intensities and thus dynamic properties. The rate at which particles diffuse is related to their size, provided all other parameters are constant [15]. DLS is accurate, reliable, noninvasive, and consumes very less time in the analysis of nanoparticles. Furthermore, this approach allows the investigation of particles in their own environment [16]. DLS also helps in determining whether the nanoparticles are monodisperse or polydisperse. If the peak is sharp, the particles are monodisperse, but if the peak is broad, the particles are polydisperse.

3.5.1.1   Principle

Particles suspended in a liquid undergo Brownian motion. The larger the particle, the slower the Brownian motion. DLS monitors the Brownian motion with light scattering. This technique measures the time-dependent fluctuations in the scattering intensity to determine the diffusion coefficient (D) and, subsequently, the hydrodynamic size.

3.5.1.2   Parameters study through DLS

Size and zeta potential of nanoparticles, polydispersity (Pd is representative of the particle size distribution width) can be studied through DLS.

3.5.1.3   Disadvantages

Only liquid samples are employed, and signal from larger particles dominates over that of smaller ones.

3.5.2 X-Ray Diffraction

Solid matter can be classified as amorphous and crystalline. In amorphous solid materials, atoms are arranged in a random or disordered way, while in the case of crystalline substances, the atoms are arranged in a regular pattern. This smallest volume element is called a unit cell. The dimensions of the unit cell are described by three axes—*a*, *b*, and c—and the angles between them—alpha, beta, and gamma. About 95% of all solids can be described as crystalline. When an X-ray beam hits an atom, the electrons around the atom start to oscillate with the same frequency as the incoming beam. The scattering of X-rays from atoms produces a diffraction pattern, which contains information about the atomic arrangement within the crystal [17]. The crystal structure of the nanomaterial reveals the atomic arrangement of the material; if the atoms are arranged differently, different diffractions are produced. The diffraction peaks produced can be linked to planes of atoms to aid in studying the atomic structure and microstructure of a sample [18, 19].

3.5.2.1 Principle

In X-ray diffraction (XRD), X-rays are passed through a crystalline material and the patterns produced give information about the size and shape of the unit cell. X-rays passing through a crystal bent at various angles: This process is called diffraction. In a nutshell, X-rays interact with electrons in matter, i.e., are scattered by the electron clouds of atoms.

3.5.2.2 Parameters study through XRD

Through this technique, identification of a material and its composition, crystallinity, and lattice parameters is possible. (If the peak is sharp and narrow, the compound is crystalline; if the peak is very broad, the material is amorphous.)

3.5.2.3 Disadvantages

This technique is time consuming and requires a large volume of sample.

3.6 Spectroscopy Methods

The interaction between nanoparticles and the electromagnetic radiation produces characteristic bands. These characteristic bands produced give information about the composition, size, and size distribution of nanoparticles. The sources of light are different according to the instrument; laser, X-rays, or neutrons can be used.

3.6.1 Fourier Transform Infrared Spectroscopy

The Fourier transform spectrometer records in the mid-infrared region (400–4500 cm^{-1}). Fourier transform infrared spectroscopy (FTIR) has appeared as an important tool for understanding the association of surface functional biological groups in metal interactions. Due to the complex interaction of atoms within the molecule, infrared absorption of the functional groups may vary over a wide range. However, it has been found that many functional groups give characteristic infrared absorption at specific narrow frequency range. Multiple functional groups may absorb at a particular frequency range, but a functional group often gives rise to several characteristic absorptions. Thus, the spectral interpretations should not be confined to one or two bands only; actually the whole spectrum should be examined [6, 7].

3.6.1.1 Principle

In this technique, nanoparticles interact with the electromagnetic radiation, producing characteristic bands. A sample is placed between a light source and a photodetector, and the intensity of a beam of light is measured before and after it passes through the sample.

3.6.1.2 Parameters study through FTIR

Identification of the presence of certain functional groups, detection of a pure compound, or verification of the presence of specific impurities can be done through this technique.

3.6.1.3   Disadvantage

FTIR instruments have only single beam while dispersive instruments generally have a double beam.

3.6.2   UV/Vis Spectroscopy

Nanoparticles have distinct optical properties, which vary according to size, shape, concentration, and refractive index near the nanoparticle surface. By exploiting these properties, we can identify and characterize the materials. When the sample does not absorb, it shows transmittance, which is close to 100% in the near infrared. In the UV portion of the spectrum where the sample absorbs strongly, the transmittance drops to around 10% or less. It can be used as a simple and reliable method for monitoring the stability of nanoparticle solutions. If the particles destabilize, the original extinction peak will decrease in intensity, and if the nanoparticles aggregate, the peak will broaden or a secondary peak will form at longer wavelengths.

3.6.2.1   Principle

In this technique, interaction between nanoparticles and the electromagnetic radiation occurs, which produces characteristic bands. A sample is placed between a light source and a photodetector, and the intensity of a beam of light is measured before and after it passes through the sample.

3.6.2.2   Parameters study through UV-VIS

This method can be utilized to identify a pure compound or detect the presence of specific impurities, nanoparticles aggregation.

3.6.2.3   Disadvantages

This technique is not very sensitive, and structure information can be obtained only for sizes greater than 300 nm.

3.7   Conclusion

Nanomaterials show numerous features such as high surface area and promote electrical conductivity, which make these materials striking for various applications. This chapter describes various characterization techniques used for the analysis of size, composition, and morphology of nanomaterials. Different types of techniques have been used for characterizing various nanomaterials, such as TEM, SEM, UV, and XRD. Scientists use these techniques for verifying their research. This chapter emphasizes on the principle, parameters, and disadvantages of various characterization techniques.

Problems

1. How can we avoid the aggregation of nanoparticles in SEM images?
2. Why there are no sharp peaks in the XRD of nanoparticles?
3. Why there is a cloudy appearance of nanoparticles in SEM imaging?
4. How can we get a fully dispersed suspension of nanoparticles for DLS and zeta potential measurement?
5. What is the difference between SEM and TEM techniques?
6. What are the advantages of SEM and AFM?
7. How can we analyze an SEM image? What all data can be inferred from SEM?
8. How the AFM technique is better than other microscopic techniques?
9. How do I prepare a powder sample for AFM?
10. How do I use the AFM for analyzing nanomaterials?

References

1. Domingos, R.F., Baalousha, M.A., Ju-Nam, Y., Reid, M.M., Tufenkji, N., Lead, J.R., Leppard, G.G., and Wilkinson, K.J. (2009). Characterizing manufactured nanoparticles in the environment: Multimethod determination of particle sizes, *Environ. Sci. Technol.*, **43**, pp. 7277.
2. Richard, C., Jr. Charles, A., and Wilson, S. (1992). *Encyclopedia of Materials Characterization: Surfaces, Interfaces, Thin Films* (Stoneham, Massachusetts: Butterworth-Heinemann, Elsevier).
3. Joshia, M., Bhattacharyya, A., and Wazed, A.S. (2008). Characterization techniques for nanotechnology applications in textiles, *Indian J. Fibre Textile Res.*, **33**, pp. 304–317.

4. Henry, C.R. (2005). Morphology of supported nanoparticles, *Prog. Surf. Sci.*, **80(3–4)**, pp. 92–116.

5. Brydson, R., and Brown, A. (2008). An investigation of the surface structure of nanoparticulate systems using analytical electron microscopes corrected for spherical aberration. In: *Turning Points in Solid-State, Material and Surface Science* (London: RSC Publishing), pp. 778–791.

6. Howe, J.M., Mori, H., and Wang, Z.L. (2008). In situ high-resolution transmission electron microscopy in the study of nanomaterials and properties, *MRS Bull.*, **33(2)**, pp. 115–121.

7. Hutchison, J.L., Titchmarsh, J.M., Cockayne, D.J.H. et al. (2002). ACs corrected HRTEM: Initial applications in materials science, *JEOL News*, **2**, pp. 37.

8. Yamasaki, J., Kawai, T., Kondo, Y., et al. (2008). A practical solution for eliminating artificial image contrast in aberration-corrected TEM, *Microsc. Microanal.*, **14**, pp. 27–35.

9. Wang, Z.L. (1998). Structural analysis of self-assembling nanocrystal superlattices, *Advan. Mater.*, **10**, pp. 13–30.

10. Balnois, E., Papastavrou, G., and Wilkinson, K.J. (2006). Force microscopy and force measurements of environmental colloids. In: *Environmental Colloids and Particles: Behaviour, Separation and Characterisation*, Vol. 10, Wilkinson, K.J., and Lead, J.R. (eds.) (Chichester, UK: John Wiley & Sons, Ltd), pp. 405.

11. Storey, R.A., and Ymen, I. (eds)(2011). *Solid State Characterization of Pharmaceuticals* (Chichester, UK: John Wiley & Sons, Ltd).

12. Hosokawa, M., Nogi, K., Naito, M., and Yokoyama, T. (eds)(2008). *Nanoparticle Technology Handbook* (Amsterdam: Elsevier).

13. Zhong, Q., Inniss, D., Kjoller, K., and Elings, V.B. (1993). Fractured polymer/silica fiber surface studied by tapping mode atomic force microscopy, *Surf. Sci. Lett.*, **290**, pp. 688–692.

14. Keller, D., and Bustamante, C. (1993). Attaching molecules to surfaces for scanning probe microscopy, *Biophys. J.*, **64**, pp. 896.

15. Berne, B.J., and Pecora, R. (2000). *Dynamic Light Scattering: With Applications to Chemistry, Biology, and Physics* (New York: Dover Publications, Inc.).

16. Chu, B. (1992). *Laser Light Scattering: Basic Principles and Practice*, 2nd Edn (Boston: Academic Press).

17. Dann, S.E. (2002). *Reactions and Characterization of Solids* (Cambridge, UK: Royal Society of Chemistry), pp. 10.

18. Skoog, D.A., Holler, F.J., and Crouch, S.R. (2007). *Principles of Instrumental Analysis*, 6th Edn (Belmont, California: Thomson Brooks/Cole).

19. Skakle, J. (2005). Application of X-ray powder diffraction in material chemistry, *Chem. Recources,* **5**, pp. 252–262.

Chapter 4

Fabrication of Sensors for Electrochemical Determination

Jagriti Narang,[a] **Nitesh Malhotra,**[b] **and Chandra Shekhar Pundir**[c]

[a]*Amity Institute of Nanotechnology, Amity University, Noida-201313, India*
[b]*Amity Institute of Physiotherapy, Amity University, Noida-201313, India*
[c]*Department of Biochemistry, MD University, Rohtak-124001, India*
jags_biotech@yahoo.co.in

4.1 Introduction

Analysis of serum metabolites, industrial compounds, drugs, and environmental pollutants is very important because of their impact on human health. Various methods are already available for the detection of these compounds, but these methods have some drawbacks, such as time-consuming sample preparation, costly instruments, tedious procedures, and infeasibility for onsite monitoring. Analysis of biological materials is expensive and has to be performed in outside laboratories equipped with more sophisticated instrumentation. Therefore, there is a need for simple, fast, sensitive, and specific methods. The ability of isolating and purifying the proteins/

Biosensors: An Introductory Textbook
Edited by Jagriti Narang and Chandra Shekhar Pundir
Copyright © 2017 Pan Stanford Publishing Pte. Ltd.
ISBN 978-981-4745-94-9 (Hardcover), 978-1-315-15652-1 (eBook)
www.panstanford.com

enzymes and other biological elements such as cells or organelles has allowed their integration with physicochemical transduction devices to produce biosensors [1]. Researchers have now given more attention to develop various biosensors for their biomedical, environmental, and industrial applications.

Biosensors can be classified according to their biological element or transduction element. Biological elements can be enzymes, antibodies, microorganisms, biological tissues, and cell organelles. Transduction elements depend on the type of physicochemical change resulting from the sensing event. It can be in the form of current, voltage, heat, etc. [1]. When there is an interaction between the biological component and the analyte, a detectable change is produced, whose magnitude is proportional to the concentration of the analyte [2]. In this chapter, we will focus on electrochemical biosensors based on various bioreceptors such as enzymes, DNA, and antibodies.

4.2 Enzyme-Based Electrochemical Biosensors

Enzyme-based electrochemical biosensors have been exploited in various applications such as in biomedicine, food analysis, and environmental monitoring. In biomedical applications, these sensors are employed for the determination of various serum metabolites such as bilirubin [3], hydrogen peroxide [4], polyphenol [5], and triglycerides [6]. Enzyme-based biosensors are also employed for the analysis of food, such as xanthine and vitamins. In pharmaceutical applications, drug monitoring also involves enzyme-based biosensors such as deferiprone [7] and paracetamol [8]. Environmental applications of enzymes for sensing include detection of pesticides and phenolic compounds [9]. The first step in fabricating enzymatic sensors involves preparation of a sensing interface. Nanomaterials play an important role in the preparation of a sensing interface. Therefore, various nanomaterials are employed for the immobilization of various bioreceptors. Nanomaterials show some distinct properties, such as fast electron-transfer kinetics, large surface area for the immobilization of bioreceptors, and biocompatible environment. Nanomaterials are deposited on the substrate/electrode by various methods such as dip coating and electro-polymerization.

- Electrochemical deposition of metal nanoparticles on electrode is usually carried out in a solution containing a precursor of metal salts. The solution can be either acidic or basic depending on the precursor. The solution is added to the electrochemical cell and is connected by a potentiostat. In three electrode system, metal gets deposited on the surface of cathode only (working electrode). The electrodeposition process for nanoparticle synthesis is accomplished by scanning between a few ranges of voltages versus Ag/AgCl, at a suitable scan rate for suitable cycles in an electrolytic bath containing precursor salts. We can employ various methods for the electrodeposition of nanoparticles, such as cyclic voltammetry, differential pulse voltammetry, and double pulse voltammetry.

- Electrophoretic deposition of nanoparticles is achieved through the transport of positively charged nanoparticles toward a negative electrode under the applied electric field. This process also helps in the deposition of charged nanoparticles to charged electrodes under the influence of an electric field [9].

- In dip coating methods, the electrode is dipped in the solution for about 5 min and then withdrawn. After the surface is coated, the electrode is put in a dryer and stored there for subsequent procedures. After storing, the electrode is further heated at an elevated temperature [10].

The second step in the fabrication of enzyme-based biosensors is the immobilization of enzymes on the surface of the electrode. For fabrication, the enzyme has to be well attached to the interface without losing enzymatic activity. The electrode with any nanoparticles should provide a biocompatible microenvironment to the enzyme to ensure its biocatalyst activities. The immobilization method depends on many parameters, such as the nature of the enzyme, electrode, nanoparticles, and analyte. The enzyme is easily affected by the ambient conditions such as pH, temperature, and substrate concentration. Therefore, these parameters should be kept in mind to avoid the loss of enzymatic activity [13]. Enzymes are immobilized by various methods such as physical adsorption, covalent linkage, cross-linking, and entrapment.

- **Physical adsorption:** It is an easy and rapid method for the preparation of enzymatic biosensors. In this approach, negatively charged nanoparticles are allowed to bind with positively charged amino acid residues. In this process, the enzyme gets immobilized on the interface, when the modified electrode is dipped into the enzyme solution.

- **Covalent linkage:** In this method, covalent bonds form between the nanoparticles/polymers and the free amino group of the enzyme. This linkage is strong as compared to adsorption. It requires mild conditions under which reactions are performed, such as low temperature and low ionic strength and pH in the physiological range [13]. For example, carboxylated multi-walled carbon nanotube (cMWCNT) is employed as an interface support for the enzyme; the free carboxyl group of the MWCNT combines with the free amino group of the enzyme to form a covalent bond through N-ethyl-N-(3-dimethylaminopropyl) carbodiimide (EDC) and N-hydroxysuccinimide (NHS) chemistry.

- **Cross-linking**: In this procedure, the enzyme is chemically bonded to the interface by a cross-linking agent such as glutaraldehyde. Cross-linking agents are mostly bifunctional and as such they form a network of bonding.

- **Entrapment**: In this process, the enzyme solution is mixed with the monomer solution and then polymerized to a gel. This procedure helps in trapping the enzyme within the gel network, e.g., polyacrylamide, agarose gels, chitosan, etc.

- **Self-assembling monolayers:** In this process, gold-modified electrodes can be prepared by covalently tethering the gold nanoparticles with the surface functional groups ($-CN$, $-NH_2$, or $-SH$) of self-assembling monolayer modified electrode surface. Short-chain molecules such as 3-mercaptoproprionic acid and cystamine can be self-assembled on the modified electrode for further nanoparticle binding [14]. These molecules also provide the functional groups necessary for the covalent immobilization of the enzyme. This modified layer forms very dense layers, which limit the diffusion, and may cause steric hindrance about the active site, which limits bioactivity [15].

After enzyme immobilization, the modified electrode with the enzyme interacts with the analyte over the applied system, which results in the change of a physical property, which can be detected and converted into an electrical signal by the physical transducer. After interaction, physical change can be in the form of current, voltage, oxygen, etc. Amperometric biosensors measure the change in current. They are based on enzymes that either consume oxygen (e.g., all oxidases) or produce hydrogen peroxide. The produced hydrogen peroxide splits into protons, oxygen, and electrons under the applied voltage. The generated electrons cause the flow of current. The overall chemical energy converts into electrical energy. Various reactions involved with amperometric biosensors are as follows:

$$\text{Substrate} + O_2 \xrightarrow{\text{Enzyme}} \text{Product} + H_2O_2$$

$$H_2O_2 \xrightarrow{\text{Applied voltage}} O_2 + 2H^+ + 2e^-$$

$$\text{Substrate} + NAD/FAD \xrightarrow{\text{Enzyme}} \text{Product} + NADH/FADH + H^+$$

$$\text{Substrate} + O_2 \xrightarrow{\text{Enzyme}} \text{Product} + H_2O$$

$$2H_2O + 4e^- + O_2 \xrightarrow{\text{Applied voltage}} 4OH^-$$

The use of mediators such as flavin adenine dinucleotide (FAD) or nicotinamide adenine dinucleotide (NAD) has been proven as an effective approach in sensing applications as mediators facilitate charge transfer between the analyte and the enzyme-modified electrode. If water is produced after the interaction of the enzyme and the substrate, then water molecules produce hydroxyl ions at the applied voltage. Oxygen consumption is directly proportional to the concentration of the analyte. Various enzymatic sensors are employed for various applications. Some enzymatic sensors with their principles are shown in Table 4.1.

Table 4.1 Different reactions involved in sensing different analytes

Analyte	Enzyme	Reaction involved	Reference
Glucose	Glucose oxidase (GOx)	β-D-Glucose + $O_2 \longrightarrow$ Gluconic acid + H_2O_2	[16]
Fructose	Fructose dehydrogenase	Fructose + Mediator $\longrightarrow$ 5-keto-D-fructose + H_2O_2	[17]
Lactose	β-galactosidase in combination with glucose oxidase or galactose oxidase	Lactose + $H_2O \longrightarrow$ D-galactose + Glucose β-D-Glucose + $O_2 \longrightarrow$ D-gluconic acid + H_2O_2	[18]
Citric acid	Citrate lyases (CL), oxaloacetate decarboxylase (OACD), and pyruvate oxidase (POD)	Citrate $\xrightarrow{\text{CL}}$ Oxaloacetate + CH_3COOH Oxaloacetate $\xrightarrow{\text{OACD}}$ Pyruvate + CO_2 Pyruvate + $HPO_2 + 4O_2 \xrightarrow{\text{POD}}$ Gluconic acid + H_2O_2 + CO_2	[19]
Glutathione	Glutathione oxidase	Glutathione + $O_2 \xrightarrow{\text{GSHO}_x}$ Glutathione disulfide + H_2O_2	[20]
Acetylthiocholine chloride	Acetylcholinesterase	Acetylthiocholine chloride + $H_2O \longrightarrow$ Thiocholine + Acetic acid + Cl^- 2Thiocholine $\longrightarrow$ Dithio-bis-choline + $2H^+ + 2e^-$	[21]
L-Ascorbic acid	Ascorbate oxidase	L-Ascorbic acid + $\frac{1}{2}O_2 \xrightarrow[\text{oxidase}]{\text{Ascorbate}}$ Dehydroascorbic acid + H_2O	[22]
Cholesterol	Cholesterol oxidase	Cholesterol + $O_2 \xrightarrow{\text{ChO}_x}$ Cholestenone + H_2O_2	[23]

Analyte	Enzyme	Reaction involved	Reference
Choline	Acetylcholinesterase and choline oxidase	Acetylcholine chloride + H_2O $\xrightarrow{\text{AchE}}$ Acetate + Choline + H^+ Choline + $2O_2$ + H_2O $\xrightarrow{\text{ChO}_X}$ Betaine + $2H_2O_2$	[24]
Polyphenols	Polyphenol oxidase	L-DOPA + $\frac{1}{2}O_2$ $\xrightarrow{\text{PPO}}$ Benzoquinone derivative + H_2O	[25]
Xanthine	Xanthine oxidase	Xanthine + O_2 + H_2O $\longrightarrow$ Uric acid + H_2O_2	[26]
Triglycerides	Lipase, Glycerol kinase (GK), Glycerol-3-phosphate oxidase (GPO)	Triglycerides $\xrightarrow{\text{Lipase}}$ Glycerol + fatty acids Glycerol + ATP $\xrightarrow[\text{kinase}]{\text{Glycerol}}$ α-Glycerol-3-phosphate + ADP Glycerol-3-phosphate + O_2 $\xrightarrow{\text{GPO}}$ Dihydroxyacetone phosphate + H_2O_2	[27–30]
Deferiprone	Horse radish peroxidase (HRP)	Deferiprone + H_2O_2 $\longrightarrow$ Corresponding dione	[31, 32]
Paracetamol		Paracetamol + H_2O_2 $\xrightarrow{\text{HRP}}$ Acetyl-p-benzo quinoneimide	[33, 34]
Leviteracetum	Horse radish peroxidase (HRP)	$LEV_{(ox)}$ + H_2O_2 $\xrightarrow{\text{HRP}}$ $LEV_{(red)}$	[35, 36]

4.3 DNA-Based Sensors

The first DNA-based biosensor was investigated by Millan and Mikkelsen in 1993 [37]. A DNA biosensor is based on the principle of immobilization of a single-stranded DNA (ssDNA) on an interface, which combines with its complementary sequence according to Chargaff's rule and form a hybrid. After the interaction of the probe and its complementary sequence, some detectable change is produced, which can be in the form of current [38], light [39], and resonance [40]. DNA and immuno-based biosensors bind specifically to their target analytes. DNA-based biosensors offer many advantages over other biosensors since interactions between complementary sequences are very specific and robust. Different methods are employed in DNA-based biosensors for the detection of target analytes, such as direct electrochemistry of DNA, redox-active species as labels for hybridization detection, and electrochemical amplifications with nanoparticles.

4.3.1 Immobilization Techniques Used for Developing DNA Biosensors

For the preparation of sensitive DNA-based biosensors, the ssDNA should be properly immobilized on the interface or electrode [41]. Immobilization of DNA is a very important step in the fabrication of DNA-based biosensors. DNA probes are immobilized through various processes such as physical adsorption, electrodeposition, covalent immobilization, cross-linking, and avidin/streptavidin–biotin interaction [42, 43]. In this section, we discuss the various methods used for immobilizing a DNA probe to develop a DNA-hybridized sensor.

4.3.1.1 Adsorption

Adsorption is an easy and rapid method for the preparation of DNA-based biosensors [44]. In this approach, positively charged nanoparticles or polymers are allowed to bind with the negatively charged DNA molecule [45]. Researchers have exploited chitosan polymer for the immobilization of DNA probe [45, 46]. They fabricated a disposable DNA sensor by physically adsorbing amplified human cytomegalovirus DNA onto a screen-printed electrode (SPE) [47].

Procedure: Researchers have also immobilized the DNA probe using the dip coating method. The DNA probe was added drop wise to the interface and was air dried overnight. It was then soaked in water for 2 h and again rinsed with water to remove unbound DNA. Consequently, an ssDNA-modified electrode was obtained [48].

4.3.1.2 Electrodeposition

DNA probes can also be electrochemically deposited on the electrode at some specific potential [49]. For electrodeposition, very less quantity of DNA is required. This technique is fast, robust, and does not require DNA or substrate functionalization. Moreover, we can carry out electrodeposition on the nanoparticle-modified electrode.

Procedure: Researchers have electrodeposited ssDNA probes on a pretreated electrode by applying a potential of +0.50 V for 5 min in a 10 ppm probe solution containing 20 mM NaCl with 200 rpm string [49].

4.3.1.3 Covalent immobilization

In the covalent immobilization method, covalent bonds form between nanoparticles/polymers and the functionalized DNA molecule ($-NH_2$-DNA or SH-DNA). This linkage is strong compared to adsorption [50–52]. Covalent immobilization offers strong bonding, stability, and structural flexibility and promotes DNA hybridization [53]. The $-NH_2$ bond of the modified DNA probe covalently attaches to the free carboxyl group of the sensing interface modified with polyaniline, poly(3-pyrrolylacrylic acid), poly(5-(3-pyrrolyl)2,4-pentadienoic acid), and poly(3-pyrrolylpentanoic acid) [54]. Researchers have exploited functionalized nanomaterials or polymers such as $-COOH$-MWCNT and SH-AuNPs for the covalent attachment of DNA probe [55–57].

4.3.1.4 Cross-linking

In the cross-linking procedure, the $-NH_2$-modified DNA probe is chemically bonded to the interface by a cross-linking agent such as glutaraldehyde. Cross-linking agents are mostly bifunctional, because of which they form a network of bonding. In this approach, the first step involves treatment of the electrode with the polymer terminating with the $-NH_2$ group, such as chitosan and polyaniline.

Then the free $-NH_2$ group combines with the free aldehyde group ($-CHO$ group) of glutaraldehyde, while another free $-CHO$ group combines with the $-NH_2$ group of the modified DNA probe. By this approach, an NH_2-substituted DNA probe can be cross-linked onto the surface of the polymer-modified electrode [58].

4.3.1.5 Avidin/streptavidin–biotin interaction

Avidins are large protein molecules having four identical sites for binding biotin [59]. The strong binding capability of avidin and biotin has been exploited by various researchers for attaching the DNA molecule on the sensing interface [60]. Several strategies have been developed for attaching biotin to modified electrodes, including the biotin-sandwich technique for immobilizing DNA [61–63].

4.3.2 Principle of Direct Electrochemistry of DNA

Direct electrochemistry of DNA depends on the intrinsic electrochemical behavior of bases [64]. However, deoxysugar present in the DNA has also been reported to be oxidized at a copper microelectrode [65]. Bases such as adenine and guanine have been reported to be electrochemically reduced at a mercury electrode or oxidized at silver, gold, carbon, copper, etc. Adenine and guanine oxidation signals on carbon electrodes can be observed at around 1.0 V and 1.3 V in 0.50 M acetate buffer solution (pH 4.80, ABS), respectively, as reported by Jelen et al. [66]. After hybridization with the complementary sequence, the sensing signal in the form of oxidation and reduction peaks gets decreased [66–68]. The sensing signals decrease after hybridization with complementary sequences since free adenine and guanine bases are not available. Recently, researchers have exploited the affinity between avidin and biotin. An avidin-modified electrode can effectively bind to biotin-modified oligonucleotides onto the electrode surface and vice versa [69].

4.3.3 Principle of Redox-Active Label or Indirect Detection of DNA

The indirect method of DNA detection is considerably more admired than the label-free approaches because there are many more ways in which transduction can be configured with high sensitivity and selectivity [70]. In this approach, labels should be typically redox-active molecules [71–73] having different affinity toward DNA. They can either intercalate between the base pairs of a DNA duplex or interact with ssDNA. Redox mediators are small size compounds that enable the reversible exchange of electrons with the electrode. The most used electron mediators are ferrocene, $K_3Fe[(CN)]_6^{3-/4-}$, $Ru(bpy)_3^{3+/2+}$, and methylene blue.

The probe-modified electrode is immersed in a solution containing a redox-active and DNA-specific binding molecule. The DNA-specific binding molecule can specifically bind to the ssDNA or double-stranded DNA (dsDNA) according to the nature of the binding agent; for example methylene blue binds to ssDNA. The sensing signal decreases after hybridization through the electrochemical technique [74–76]. In the case of drug interactions, it has been shown that this intercalating agent, i.e., daunomycin (DM), shifts the oxidizing potential [77]. Researchers have reported that the anodic peak potentials of the drug shift to more positive values in the case of a dsDNA-modified electrode compared with an ssDNA-modified one.

Another mediator is ferrocene, which is tagged with the DNA stem loop structure and immobilized on the electrode. The tagged ferrocene remains in close contact with the electrode due to its stem loop structure, which hybridizes itself. When the tagged ferrocene is in close contact with the electrode, an amplified sensing signal is produced. Once hybridization occurs, the sensing signal decreases as the tagged ferrocene moves apart from the electrode surface, hence switching off the electrochemical signal [78, 79].

Enzyme-labeled DNA is also employed for the detection of DNA. For instance, glucose oxidase can be used for the detection of viral genes on an Au electrode [80]. In this approach, researchers have used ferrocene-conjugated uracil bases to generate a redox-active compound. The generated compound acts as an electron-transfer mediator for the bioelectrocatalysis of glucose.

4.4 Immunosensors

Electrochemical immunosensors are the best among all sensors because of their high sensitivity, ease of fabrication, and specificity.

The principle behind electrochemical immunosensors is the exceptional specificity of antigen and antibodies, which is the basis for a detectable change [81]. The incorporation of nanomaterials in the fabrication of sensors offers many advantages with potentially improved performance [82]. Because of the high specificity between antigen and antibody, most of the immunosensors are not reusable. There is a great apprehension in the regeneration of immunosensors, which restricts its ability of reuse. In some immunosensors, researchers also employ aptamers instead of antigen and antibody. Aptamers are ssDNA molecules, i.e., synthetic antibodies, which specifically bind to molecular targets, such as proteins. An interesting aspect of using aptamers as the recognition element in place of antibodies is that it makes possible label-free biosensors [83]. Synthetic antibodies are immobilized onto the sensing interface from one end and linked to a redox label from the other end. When a bond is formed between the aptamer and molecular targets, it results in either activation or deactivation of the redox probe, which leads to a change in redox activity.

4.4.1 Immobilization Techniques Used for Developing Immunosensors

Immobilization is an important parameter in the fabrication of immunosensors. Antibodies have to be properly immobilized on the immunosensor surface. Proper orientation and homogeneous distribution on the sensing interface are of paramount importance. If an antibody is not immobilized on the sensing interface with proper orientation, it may lose its biological activity and cause inactivation. Therefore, in the case of immunosensor, the immobilization method is of utmost importance. The antibody should be immobilized with proper orientation so that it can specifically bind to the antigen with minimal steric hindrance for the fabrication of immunosensors [84]. Different approaches are employed for immobilizing antibodies on the sensing interface, which includes physical adsorption, covalent attachment, entrapment within a polymer matrix, and directed immobilization.

4.4.1.1   Physical adsorption

Physical adsorption is the simplest method of binding an antigen/ antibody on a surface. It is an easy and rapid method for the preparation of immunosensors. In this process, the antibody is immobilized on the interface by dipping the modified electrode into the antibody solution. It can occur through hydrophilic, hydrophobic, or both types of interactions between antibodies and the sensor surface [85]. Furthermore, these surfaces may be modified to introduce some hydrophilic or hydrophobic groups, which help in the interaction of antibodies and the sensing interface. The hydrophobic interactions that drive the passive adsorption of antibodies can reportedly reduce the number of functional sites or activities by more than 90% [86–88]. The main disadvantage is desorption of antibodies from the sensing interface.

4.4.1.2   Entrapment

Entrapment is a non-covalent immobilization method of antibodies, which involves conducting polymer films such as polyaniline and polyindole-5-carboxylic acid. In this method, antibodies are entrapped in between the monomeric units of polymer on the sensing interface. During the preparation of an antibody-entrapped polymer matrix, antibodies can denature and also become inaccessible because they are buried deep within the polymer film.

4.4.1.3   Electrodeposition

To overcome the problem of inaccessibility and denaturation, the technique of electrodeposition is employed. In this approach, antibodies are electro-polymerized along with monomers by applying a potential to the electrode surface, which enhances electrostatic interactions with the polymer backbone and the antibody, thereby, facilitating immobilization [88].

4.4.1.4   Covalent attachment

In covalent attachment, a covalent bond is formed between the nanoparticles/polymers and the antibodies. The main advantage of this approach is that it ensures adequate immobilization exclusive of desorption of the biomolecule from the sensing interface. Covalent immobilization involves coupling of antibodies through free amino groups to a carboxyl-activated sensor surface. Various groups can be introduced on the sensing interface, such as carboxylic acids (–COOH), aldehyde group (–CHO), sulfhydryl group (–SH), and hydroxyl group (–OH), through various chemical compounds such as EDC (carbodiimides) and N-hydroxysuccinimide (NHS), glutaraldehyde, periodate oxidation, isothiocyanates, epoxides, aldehydes, and cyanogen bromide.

4.4.1.5 Avidin/streptavidin–biotin interaction

Avidins are large protein molecules having four identical sites for binding biotin. A commonly used affinity-based method for orienting antibodies specifically onto surfaces involves attachment of biotinylated antibodies onto a (strept)avidin-modified surface. Several strategies have been developed for attaching biotin to modified electrodes, including the biotin-sandwich technique for immobilizing antibodies.

4.5   Conclusion

Nanomaterials can be designed with the biorecognition property for sensing applications. Nanosensors based on various biomolecules, such as enzymes, DNA, and antigen–antibody, are exploited for clinical, environmental, and pharmaceutical analyses. Nanomaterials provide advantages conferred by their nanodimension such as surface area, electron conductivity, stability, and biocompatibility. Furthermore, nanomaterials can be modified for better attachment of biomolecules. Nanomaterials also show electrocatalytic properties, which help in producing redox signals of electroactive species. Additional nanomaterials provide protective ambient conditions for biomolecules, which help in protecting their physicochemical and biological attributes. Nanosensors exhibit advantageous features, for instance amplified sensing signal, stability, and even promote redox reactions. This chapter describes various methods of attachment of biomolecules on the sensing interface, principles behind sensing, and reaction involved for the determination of various serum metabolites.

Problems

1. Define the dynamic range of a biosensor?
2. What parameters should be evaluated to check the performance of a biosensor?
3. How do we calculate the Michaelis–Menten constant for amperometric enzyme-based biosensors?
4. What is the importance of using immobilized enzymes?
5. What are baseline stability, response time, and recovery time in sensor criteria?
6. How to separate antigen–antibody complexes without denaturing the proteins?
7. Why immunosensors are preferred to ELISA?
8. How can you calculate the active area of an electrode in a biosensor?

References

1. Turner, A.P.F., Karube, I., and Wilson, G.S. (1987). *Biosensors: Fundamentals and Applications* (Oxford University Press, Oxford), pp. 770.

2. Turner, A.P., Chen, B., and Piletsky, S.A. (1999). In vitro diagnostics in diabetes: Meeting the challenge, *Clin. Chem.*, **45**, pp. 1596–1601.

3. Narang, J., Malhotra, N., Mathur, A., Vivek, and Pundir, C.S. (2015) Fabrication of bilirubin biosensor by taking gold nanomaterial as a sensing interface, *Adv. Mater. Lett.*, **6**, pp. 1012–1017.

4. Narang, J., Chauhan, N., Singh, A., and Pundir, C.S. (2011). A non-enzymatic sensor for hydrogen peroxide based on polyaniline, multiwalled carbon nanotubes and gold nanoparticles modified Au electrode, *Analyst*, **136**, pp. 4460.

5. Narang, J., Chawla, S., Chauhan, N., Dahiya, M., and Pundir, C.S. (2012). Construction of an amperometric polyphenol biosensor based on PVA membrane, *Sens. Instrum. Food Qual. Saf.*, **7**, pp. 22–28.

6. Narang, J., Chauhan, N., Rani, P., and Pundir, C.S. (2013). Construction of an amperometric TG biosensor based on AuPPy nanocomposite and poly (indole-5-carboxylic acid) modified Au electrode, *Bioproces. Biosys. Eng.*, **36**, pp. 425–432.

7. Narang, J., Malhotra, N., Singh, G., and Pundir, C.S. (2015). Electrochemical impedimetric detection of anti-HIV drug taking Gold nanorods as a sensing interface, *Biosens. Bioelectron.,* **66**, pp. 332–337.

8. Narang, J., Malhotra, N., Singh, S., Singh, G., and Pundir, C.S. (2015). Monitoring analgesic drug using sensing method based on nanocomposite, *RSC Adv.,* **5**, pp. 2396–2404.

9. Narang, J., Chauhan, N., Singh, A., and Pundir, C.S. (2011). A nylon membrane based amperometric biosensor for polyphenol determination, *J. Mol. Catal. B: Enzymatic*, **72**, pp. 276– 281.

10. Putzbach, W., and Ronkainen, N.J. (2013). Immobilization techniques in the fabrication of nanomaterial-based electrochemical biosensors: A review, *Sensors,* **13**, pp. 4811–4840. doi:10.3390/s130404811.

11. Yaldagard, M., Jahanshahi, M., and Seghatoleslami, N. (2013). Carbonaceous nanostructured support materials for low temperature fuel cell electrocatalysts: A review, *World J. Nano Sci. Eng.,* **3**, pp. 121–153.

12. Sakka, S. (ed.)(2005). *Handbook of Sol-Gel Science and Technology* (Springer US, New York), pp. 1968.

13. Zhao, Z., and Jiang, H. (2010). Enzyme-based electrochemical biosensors. In: Serra, P.A. (ed.), *Biosensors* (InTECH, Rijeka).

14. Kumar, C.S.S.R. (2007). Carbon nanotube-based sensor. In: *Nanomaterials for Biosensors* (Wiley-VCH, Weinheim, Germany), pp. 27–89.

15. Challa, S.S.R.K. (2010). Gold nanocomposite biosensors. In: *Nanocomposites: Nanomaterials for the Life Sciences* (Wiley-VCH, Weinheim, Germany), pp. 139–162.

16. Zhai, D., Liu, B., Shi, Y., Pan, L., Wang, Y., Li, W., Zhang, R., and Yu, G. (2013). Highly sensitive glucose sensor based on Pt nanoparticle/ polyaniline hydrogel heterostructures, *ACS Nano*, **7 (4)**, pp. 3540–3546.

17. Ikeda, T., Matsushita, F., and Senda, M. (1991). Amperometric fructose sensor based on direct bioelectrocatalysis, *Biosens Bioelectron*, **6 (4)**, pp. 299–304.

18. Tkác, J., Sturdík, E., and Gemeiner, P. (2000). Novel glucose non-interference biosensor for lactose detection based on galactose oxidase-peroxidase with and without co-immobilised beta-galactosidase, *Analyst*, **125(7)**, pp. 1285–1289.

19. Prodromidisa, M.I., Tzouwara-Karayannia, S.M., Karayannis, M.I., and Vadgamab, P.M. (1997). Bioelectrochemical determination of citric

acid in real samples using a fully automated flow injection manifold, *Analyst*, **122**, pp. 1101–1106.

20. Narang, J., Chauhan, N., Jain, P., and Pundir, C.S. (2012). Silver nanoparticles/multiwalled carbon nanotubes/polyaniline for amperometric glutathione biosensor, *Int. J. Biol. Macromol.*, **50**, pp. 672–678.

21. Chauhan, N., Narang, J., and Pundir, C.S. (2011). Immobilization of rat brain acetylcholinesterase on porous gold-nanoparticle–CaCO$_3$ hybrid material modified Au electrode for detection of organophosphorous insecticides, *Int. J. Biol. Macromol.*, **49**, pp. 923–929.

22. Chauhan, N., Narang, J., and Pundir, C.S. (2011). Fabrication of MWCNT/ PANI modified Au electrode for ascorbic acid determination, *Analyst*, **136 (9)**, pp. 1938–1945.

23. Pundir, C.S., Narang, J., Chauhan, N., Sharma, P., and Sharma, R. (2012). An amperometric cholesterol biosensor based on epoxy resin membrane bound cholesterol oxidase, *Indian J. Med. Res.*, **136**, pp. 78–85.

24. Pundir, S., Chauhan, N., Narang, J., and Pundir, C.S. (2012). Amperometric choline biosensor based on multiwalled carbon nanotubes/zirconium oxide electrodeposited on modified glassy carbon electrode, *Anal Biochem.*, **427(1)**, pp. 26–32.

25. Narang, J., Chauhan, N., Singh, A., and Pundir, C.S. (2011). A nylon membrane based amperometric biosensor for polyphenol determination, *J. Mol. Catal. B: Enzymatic*, **72**, pp. 276– 281.

26. Devi, R., Narang, J., Yadav, S., and Pundir, C.S. (2011). Amperometric determination of xanthine in tea, coffee and fish with pencil graphite rod bound xanthine oxidase, *J. Anal. Chem.*, **67**, pp. 273–277.

27. Narang, J., Minakshi, B.M., and Pundir, C.S. (2010). Determination of serum triglyceride by enzyme electrode using covalently immobilized enzyme on egg shell membrane, *Int. J. Biol. Macromol.*, **47**, pp. 691–695.

28. Narang, J., Minakshi, B.M., and Pundir, C.S. (2010). Fabrication of an amperometric TG biosensor based on PVC membrane, *Anal. Lett.*, **43**, pp. 1–10.

29. Narang, J., and Pundir, C.S. (2011). Construction of a triglyceride amperometric biosensor based on chitosan-ZnO nanocomposite film, *Int. J. Biol. Macromol.*, **49**, pp. 707–715.

30. Narang, J., and Pundir, C.S. (2013). Determination of triglycerides with special emphasis on biosensors: A review, *Int. J. Biol. Macromol.*, **61**, pp. 379–389.

31. Narang, J., Malhotra, N., Singh, G., and Pundir, C.S. (2015). Electrochemical impedimetric detection of anti-HIV drug taking gold nanorods as a sensing interface, *Biosens. Bioelectron.*, **66**, pp. 332–337.

32. Narang, J., Malhotra, N., Singh, G., and Pundir, C.S. (2015). Voltammetric detection of anti-HIV replication drug based on novel nanocomposite gold-nanoparticle–$CaCO_3$ hybrid material, *Bioprocess Biosyst. Eng.*, **38(5)**, pp. 815–822.

33. Narang, J., Malhotra, N., Singh, S., Singh, G., and Pundir, C.S. (2015). Monitoring analgesic drug using sensing method based on nanocomposite, *RSC Adv.*, **5**, pp. 2396–2404.

34. Narang, J., Malhotra, N., Chauhan, N., Singh, G., and Pundir, C.S. (2015). Development and validation of biosensing method for acetaminophen drug monitoring, *Adv. Mat. Lett.*, **6**, pp. 209–216.

35. Narang, J., Malhotra, N., and Pundir, C.S. (2015). Multiwalled carbon nanotubes wrapped nanoflakes graphene composite for sensitive biosensing of leviteracetum, *RSC Adv.*, **5**, pp. 13462–13469.

36. Narang, J., Malhotra, N., Chauhan, N., Singh, G., and Pundir, C.S. (2015). Nanocrystals of zeolite act as enhanced sensing interface for biosensing of leviteracetum, *J. Pharma. Sci.*, 104, pp. 1153–1159.

37. Millan, K.M., and Mikkelsen, S.R. (1993). Sequence-selective biosensor for DNA based on electroactive hybridization indicators, *Anal. Chem.*, **65**, pp. 2317–2323.

38. Millan, K.M., Saraullo, S., and Mikkelsen, S.R. (1994). Voltammetric DNA biosensor for cystic fibrosis based on a modified carbon paste electrode, *Anal. Chem.*, **66**, pp. 2943–2948.

39. Piunno, P.A.E., Krull, U.J., Hudson, R.H.E., Damha, M.J., and Cohen, H. (1994). Fiber optic biosensor for fluorometric detection of DNA hybridization, *Anal. Chim. Acta,* **288**, pp. 205–214.

40. Sawata, S., Kai, E., Ikebukuro, K., Iida, T., Honda, T., and Karube, I. (1999). Application of peptide nucleic acid to the direct detection of deoxyribonucleic acid amplified by polymerase chain reaction, *Biosens. Bioelectron.*, **14**, pp. 397–404.

41. Sassolas, A., Leca-Bouvier, B.D., and Blum, L.J. (2008). DNA biosensors and microarrays, *Chem. Rev.*, **108**, pp. 109–139.

42. Fuentes, M., Mateo, C., García, L., Tercero, J.C., Guisán, J.M., and Fernández-Lafuente, R. (2004). Directed covalent immobilization of aminated DNA probes on aminated plates, *Biomacromolecules*, **5**, pp. 883–888.

43. Peng, H., Zhang, L., Soeller, C., and Travas-Sejdic, J. (2009). Conducting polymers for electrochemical DNA sensing, *Biomaterials*, **30**, pp. 2132–2148.

44. Pividori, M.I., Merkoci, A., and Alegret, S. (2000). Electrochemical genosensor design: Immobilisation of oligonucleotides onto transducer surfaces and detection methods, *Biosens. Bioelectron.*, **15**, pp. 291–303.

45. Kara, P., Kerman, K., Ozkan, D., Meric, B., Erdem, A., Nielsen, P.E., and Ozsoz, M. (2002). Label-free and label-based electrochemical detection of hybridization by using methylene blue and peptide nucleic acid probes at chitosan modified carbon paste electrodes, *Electroanalysis*, **14**, pp. 1685–1690.

46. Xu, C., Cai, H., He, P., and Fang, Y. (2001). Electrochemical detection of sequence-specific DNA using a DNA probe labeled with aminoferrocene and chitosan modified electrode immobilized with ssDNA, *Analyst*, **126**, pp. 62–65.

47. Azek, F., Grossiord, C., Joannes, M., Limoges, B., and Brossier, P. (2000). Hybridization assay at a disposable electrochemical biosensor for the attomole detection of amplified human cytomegalovirus DNA, *Anal. Biochem.*, **284**, pp. 107–113.

48. Kerman, K., Ozkana, D., Kara, P., Meric, B., Gooding, J.J., and Ozsoz, M. (2002). Voltammetric determination of DNA hybridization using methylene blue and self-assembled alkanethiol monolayer on gold electrodes, *Anal. Chim. Acta*, **462**, pp. 39–47.

49. Erdema, A., Kerman, K., Meric, B., Akarca, U.S., and Ozsoz, M. (2000). Novel hybridization indicator methylene blue for the electrochemical detection of short DNA sequences related to the hepatitis B virus, *Anal. Chim. Acta*, **422**, pp. 139–149.

50. Tam, P.D., Tuan, M.A., and Chien, N.D. (2007). DNA covalent attachment on conductometric biosensor for modified genetic soybean detection, *Commun. Phys.*, **17**, pp. 234–240.

51. Peng, H., Soeller, C., and Travas-Sejdic, J. (2007). Novel conducting polymers for DNA sensing, *Macromolecules*, **40**, pp. 909–914.

52. Rasmussen, S.R., Larsen, M.R., and Rasmussen, S.E. (1997). Covalent immobilization of DNA onto polystyrene microwells: The molecules are only bound at the 5' end, *Anal. Biochem.*, **198**, pp. 138–142.

53. Rahman, M.M., Li, X.B., Lopa, N.S., Ahn, S.J., and Lee, J.J. (2015). Electrochemical DNA hybridization sensors based on conducting polymers, *Sensors*, **15**, pp. 3801–3829.

54. Niveleau, A., Sage, D., Reynaud, C., Bruno, C., Legastelois, S., Thomas, V., and Dante, R. (1993). Covalent linking of haptens, proteins and nucleic acids to a modified polystyrene support, *J. Immunol. Methods,* **159**, pp. 177–187.

55. Santiago-Rodríguez, L., Sánchez-Pomales, G., and Cabrera, C.R. (2010). Single-walled carbon nanotubes modified gold electrodes as an impedimetric DNA sensor, *Electroanalysis,* **22**, pp. 399–405.

56. Gangopadhyay, R., and De, A. (2000). Conducting polymer nanocomposites: A brief overview, *Chem. Mater.,* **12**, pp. 608–622.

57. Wahab, R., Ansari, S.G., Kim, Y.S., Mohanty, T.R., Hwang, I.H., and Shin, H.S. (2009). Immobilization of DNA on nano-hydroxyapatite and their interaction with carbon nanotubes, *Synth. Met.,* **159**, pp. 238–245.

58. Prabhakar, N., Arora, K., Singh, H., and Malhotra, B.D. (2008). Polyaniline based nucleic acid Sensor, *J. Phys. Chem. B,* **112**, pp. 4808–4816.

59. Tichoniuk, M., Ligaj, M., and Filipiak, M. (2008). Application of DNA hybridization biosensor as a screening method for the detection of genetically modified food components, *Sensors,* **8**, pp. 2118–2135.

60. Pan, S.L., and Rothberg, L. (2005). Chemical control of electrode functionalization for detection of DNA hybridization by electrochemical impedance spectroscopy, *Langmuir,* **21**, pp. 1022–1027.

61. Dupont-Filliard, A., Billon, M., Livache, T., and Guillerez, S. (2004). Biotin/avidin system for the generation of fully renewable DNA sensor based on biotinylated polypyrrole film, *Anal. Chim. Acta,* **515**, pp. 271–277.

62. Calvo-Muñoz, M.L., Dupont-Filliard, A., Billon, M., Guillerezb, S., Bidan, G., Marquette, C., and Blum, L. (2005). Detection of DNA hybridization by ABEI electrochemiluminescence in DNA-chip compatible assembly, *Bioelectrochemistry,* **66**, pp. 139–143.

63. Baur, J., Gondran, C., Holzinger, M., Defrancq, E., Perrot, H., and Cosnier, S. (2010). Label-free femtomolar detection of target DNA by impedimetric DNA sensor based on poly(pyrrole-nitrilotriacetic acid) film, *Anal. Chem.,* **52**, pp. 1066–1072.

64. Paleček, E. (1996). From polarography of DNA to microanalysis with nucleic acid-modified electrodes, *Electroanalysis,* **8**, pp. 7.

65. Singhal, P., and Kuhr, W.G. (1997). Sinusoidal voltammetry for the analysis of carbohydrates at copper electrodes, *Anal Chem.,* **69**, pp. 4828

66. Jelen, F., Fojta, M., and Palecek, E. (1997). Voltammetry of native double-stranded, denatured and degraded DNAs, *J. Electroanal. Chem.,* **427**, pp. 49–56.

67. Meric, B., Kerman, K., Ozkan, D., Kara, P., and Ozsoz, M. (2002). Indicator-free electrochemical DNA biosensor based on adenine and guanine signals, *Electroanalysis,* **14**, pp. 1245–1250.

68. Kara, P., Kerman, K., Ozkan, D., Meric, B., Erdem, A., Nielsen, P.E., and Ozsoz, M. (2002). Label-free and label based electrochemical detection of hybridization by using methylene blue and peptide nucleic acid probes at chitosan modified carbon paste electrodes, *Electroanalysis,* **14**, pp. 1685–1690.

69. Masarik, M., Kizek, R., Kramer, K.J., Billova, S., Brazdova, M., Vacek, J., Bailey, M., Jelen, F., and Howard, J.A. (2003). Application of avidin-biotin technology and adsorptive transfer stripping square-wave voltammetry for detection of DNA hybridization and avidin in transgenic avidin maize, *Anal. Chem.,* **75**, pp. 2663–2669.

70. Odenthal, K.J., and Gooding, J.J. (2007). An introduction to electrochemical DNA biosensors, *Analyst,* **132**, pp. 603–610.

71. Erdem, A., Kerman, K., Meric, B., Akarca, U.S., and Ozsoz, M. (2000). Novel hybridization indicator methylene blue for the electrochemical detection of short DNA sequences related to the hepatitis B DNA virus, *Anal. Chim. Acta,* **422**, pp. 139–149.

72. Johnston, D.H., Glasgow, K.C., and Thorp, H.H. (1995). Electrochemical measurement of the solvent accessibility of nucleobases using electron transfer between DNA and metal complexes, *J. Am. Chem. Soc.,* **117**, pp. 8933–8938.

73. Armistead, P.M., and Thorp, H.H. (2000). Modification of indium tin oxide electrodes with nucleic acids: Detection of attomole quantities of immobilized DNA by electrocatalysis, *Anal. Chem.,* **72**, pp. 3764–3770.

74. Erdem, A., Kerman, K., Meric, B., Akarca, U.S., and Ozsoz, M. (2000). Novel hybridization indicator methylene blue for the electrochemical detection of short DNA sequences related to the hepatitis B virus, *Anal. Chim. Acta,* **422**, pp. 139–149.

75. Meric, B., Kerman, K., Ozkan, D., Kara, P., Erensoy, S., Akarca, U.S., Mascini, M., and Ozsoz, M. (2002). Electrochemical DNA biosensor for the detection of TT and hepatitis B virus from PCR amplified real samples by using methylene blue, *Talanta,* **56**, pp. 837–846.

76. Kerman, K., Ozkan, D., Kara, P., Meric, B., Gooding, J.J., and Ozsoz, M. (2002). Voltammetric determination of DNA hybridization using methylene blue and self-assembled alkanethiol monolayer on Au electrodes, *Anal. Chim. Acta,* **462**, pp. 39–47.

77. Hashimato, K., Ito, K., and Ishimori, Y. (1994). Novel DNA sensor for electrochemical gene detection, *Anal. Chim. Acta,* **286**, pp. 219.

78. Fan, C.H., Plaxco, K.W., and Heeger, A.J. (2003). Electrochemical interrogation of conformational changes as a reagentless method for the sequence-specific detection of DNA, *Proc. Natl. Acad. Sci. U.S.A.*, **100**, pp. 9134–9137.

79. Xiao, Y., Lubin, A.A., Baker, B.R., Plaxco, K.W., and Heeger, A.J. (2006). Single-step electronic detection of femtomolar DNA by target-induced strand displacement in an electrode-bound duplex, *Proc. Natl. Acad. Sci. U.S.A.*, **103**, pp. 16677–16680.

80. Patolsky, F., Weizmann, Y., and Willner, I. (2002). Redox-active nucleic-acid replica for the amplified bioelectrocatalytic detection of viral DNA, *J. Am. Chem. Soc.*, **124**, pp. 770–772.

81. Escosura-Muniz, A., and Merkoci, A. (2010). Electrochemical detection of proteins using nanoparticles: applications to diagnostics, *Expert Op. Med. Diagn.* **4**, pp. 21.

82. Narang, J., Chauhan, N., Malhotra, N., and Pundir, C.S. (2014). Fabrication of triglyceride biosensor based on magnetic nanoparticles/zinc oxide/zinc hexacyanoferrate film: Novel immobilization matrix for electrochemical sensing, *Adv. Sci. Lett.*, **20**, pp. 1331–1336.

83. Cho, I.H., Paek, E.H., Lee, H., Kang, J.Y., Kim, T.S., and Paek, S.H. (2007). Site-directed biotinylation of antibodies for controlled immobilization on solid surfaces, *Anal. Biochem.*, **365**, pp. 14–23.

84. Viguier, C., Lynam, C., and Kennedy, R.O. (2012). Trends and perspectives in immunosensors. In: Meulenberg, E.P. (ed.), *Antibodies Applications and New Developments* (Bentham Science Publishers, Potomac, Maryland), pp. 184–208.

85. Butler, J.E., Ni, L., and Nessler, R. (1992). The physical and functional behaviour of capture antibodies adsorbed on polystyrene, *J. Immunol. Methods*, **150**, pp. 77–90.

86. Butler, J.E. (2000). Solid supports in enzyme-linked immunosorbent assay and other solid-phase immunoassays, *J. Immunol. Methods*, **22**, pp. 4–23.

87. Davies, J., Dawkes, A.C., and Haymes, A.G. (1994). A scanning tunnelling microscopy comparison of passive antibody adsorption and biotinylated antibody linkage to streptavidin on microtiter wells, *J. Immunol. Methods*, **167**, pp. 263–269.

88. Morrin, A., Ngamna, O., Killard, A.J., Moulton, S.E., Smyth, M.R., and Wallace, G.G. (2005). An amperometric enzyme biosensor fabricated from polyaniline nanoparticles, *Electroanalysis*, **17**, pp. 423–430.

Chapter 5

Electrochemical Techniques

Jagriti Narang,[a] Nitesh Malhotra,[b] and Chandra Shekhar Pundir[c]

[a]*Amity Institute of Nanotechnology, Amity University, Noida-201313, India*
[b]*Amity Institute of Physiotherapy, Amity University, Noida-201313, India*
[c]*Department of Biochemistry, MD University, Rohtak-124001, India*
jags_biotech@yahoo.co.in

5.1 Introduction

Electrochemical techniques are unavoidable in the fabrication of sensing devices. These techniques reveal redox processes to untangle reaction mechanisms and are also employed in monitoring the kinetics of electron transfer processes [1]. In addition, these techniques are also employed for the study of adsorption and crystallization phenomena at the surface of the electrode [2]. Electrochemical techniques are functional in the preparation of sensors, the principal ones being polarographic and voltammetric techniques [3]. Researchers employed these techniques due to their inherent advantages such as ease of instrumentation, high sensitivity and specificity with wide linear concentration ranges, fast response time, and low detection limit.

Biosensors: An Introductory Textbook
Edited by Jagriti Narang and Chandra Shekhar Pundir
Copyright © 2017 Pan Stanford Publishing Pte. Ltd.
ISBN 978-981-4745-94-9 (Hardcover), 978-1-315-15652-1 (eBook)
www.panstanford.com

Voltammetry is a branch of electrochemistry that was developed with the discovery of polarography in 1922 by Jaroslav Heyrovsky (received the Nobel Prize in 1959). It is a branch of science that deals with voltage and current. In this technique, a potential is applied between the two electrodes and the monitoring output parameter is the resulting current flowing through the electrochemical cell. This chapter focuses on unmodified/modified nanostructured electrodes correlated with different types of electrochemical techniques—cyclic voltammetry, differential pulse voltammetry (DPV), square-wave voltammetry (SWV), chronoamperometry, and pulsed amperometry—applied for sensing application as voltammetric/amperometric sensors.

5.2 Equipment Used in Voltammetric Experiments

The equipment used for monitoring voltammetric measurements is potentiostat, which is composed of three components: a potentiostat, a personal computer, and an electrochemical cell. The electrochemical cell is an analyte holder, in which the corresponding analyte is dissolved in an appropriate solvent and placed thereafter in an ionic electrolyte, where usually three electrodes are situated: working electrode, reference electrode, and counter electrode. These three electrodes are completely immersed in the electrochemical cell. However, in some systems, the reference electrode is placed in a separate compartment to avoid contamination and is connected to the cell via an electrolyte bridge. The most commonly used reference electrodes are the calomel electrode, whose potential is determined by the reaction

$$Hg_2Cl_2(s) + 2e^- \rightarrow 2Hg\,(l) + 2Cl^-$$

and the silver/silver chloride electrode (Ag/AgCl), whose potential is defined by the reaction

$$AgCl(s) + e^- \rightarrow Ag(s) + Cl^-$$

Working electrodes are broadly classified into two categories: (i) metal electrodes, which develop an electric potential in response to a redox reaction at the metal surface, and (ii) ion-selective electrodes, which selectively bind one type of ions to a membrane to generate an electric potential.

5.3   Cyclic Voltammetry

Cyclic voltammetry is a versatile electrochemical technique that helps in getting information about an analyte by measuring current as a function of the applied potential. It allows investigation of the mechanism of the redox process and the transport properties of a system in solution. In this system, three types of transport mechanisms are involved: (i) migration—movement of ions through the solution by electrostatic attraction to the charged electrode, (ii) convection—mechanical motion of the solution as a result of stirring or flow, and (iii) diffusion—motion of a species caused by a concentration gradient.

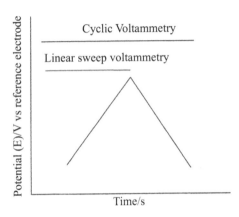

Figure 5.1   Typical voltammetry potential triangular waveform.

The potential is scanned between a higher range and a lower range: E_1 and E_2. In a cyclic voltammogram, the cycle starts from E_1 and reaches E_2 and returns to E_1. The whole cycle is called cyclic voltammogram. However, if the cycle stops at E_2 and does not return to E_1, the voltammogram obtained is called a linear sweep voltammogram (Fig. 5.1). The rate at which the whole cycle is achieved is called the voltammetric scan rate. This technique is employed for characterizing an electrode material for every type of application. A cyclic voltammetry study is performed with a three-electrode system:

 i. **Working electrode**: potential is varied with time

ii. **Reference electrode**: potential remains constant (Ag/AgCl electrode or Hg/Hg_2Cl calomel)

iii. **Counter electrode**: completes circuit; conducts e^- from signal source through the solution to the working electrode. Mostly Pt electrode is used.

The three-electrode setup is immersed in the supporting electrolyte solution, which consists of a non-reactive electrolyte to conduct current. The potential is scanned between a range of extreme values linearly with time using a triangular potential waveform.

The experiments are performed with a potentiostat instrument, which effectively controls the voltage between the reference electrode and the working electrode, while measuring the current through the counter electrode (the working electrode is connected to the ground). The electrolyte is usually added to the test solution to ensure sufficient conductivity [1]. The potential is scanned back and forth linearly with time between the two extreme values. When the potential of the working electrode is more positive than that of a redox couple present in the solution, the corresponding species may be oxidized (i.e., electrons going from the solution to the electrode) and produce an anodic current. Similarly, on the return scan, as the working electrode potential becomes more negative than the reduction potential of a redox couple, reduction (i.e., electrons flowing away from the electrode) may occur to cause a cathodic current. By the IUPAC convention, anodic currents are positive and cathodic currents are negative [2].

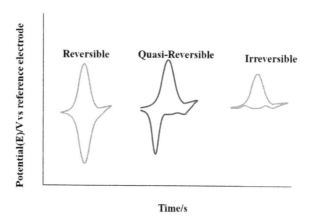

Figure 5.2 Cyclic voltammetry curves for a reversible, a quasi-reversible, and an irreversible electrode process.

For irreversible processes, peaks are reduced in size and widely separated. In completely irreversible systems, peaks are characterized by a shift in the peak potential with the scan rate. Voltammograms of a quasi-reversible system are more drawn out and exhibit a larger separation in peak potentials compared to a reversible system (Fig. 5.2).

5.3.1 Principle of Cyclic Voltammetry

The electroactive species in the sample solution are drawn toward the working electrode where a half-cell redox reaction takes place. The other corresponding half-cell redox reaction takes place at the counter electrode to complete the electron flow. A plot of current as a function of the applied potential is called a voltammogram. It is the electrochemical equivalent of a spectrum, which provides quantitative and qualitative information about the species involved in the redox reaction. The resultant current flow is linearly proportional to the concentration of the electroactive species (analyte) involved.

They are characterized by peak potentials, E_p, at which the current reaches a local maximum or minimum and the value of the peak current, i_p, at these points (Fig. 5.3).

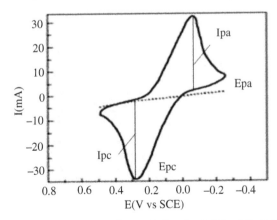

Figure 5.3 Cyclic voltammogram for a solution that is 6.0 mM in $K_3Fe(CN)_6$ and 1.0 M in KNO_3.

E_{pa}: anodic peak potential

E_{pc}: cathodic peak potential

I_{pa}: peak current at anodic potential

I_{pc}: peak current at cathodic potential

Analysis of cyclic voltammogram graph:

(i) Negative charge current value due to oxidation of water.

(ii) No observable current in this range due to the absence of reducible or oxidizable species in this potential range.

(iii–iv) Current starts attaining height at 0.3 V; oxidation reaction starts at anode. It means the oxidizable species starts decreasing, and the reduced compound starts increasing.

(v) The current decays rapidly as the diffusion layer is extended further away from the electrode surface.

(vi) Ultimately, the oxidation process stops and then the reduction process starts occurring and the cathodic current results from the reduction of the oxidized compound.

(vii) The cathodic current starts decreasing as the accumulated reduced compound starts being consumed.

5.4 Polarogram Graph

A graph of current versus potential in a polarographic experiment is called a polarogram.

- No reduction of the compound is observed as the potential is scanned slightly negative with respect to the reference electrode. A small current is observed, which is a capacitive current.

- After a certain potential is attained, reduction of the compound starts and the current increases.

- When the potential is sufficiently negative around 1.2 V, reduction of the compound begins and the curve rises steeply.

- At positive potentials, oxidation of the Hg electrode produces a negative current. By convention, a negative current means that the working electrode is behaving as the anode with respect to the auxiliary electrode. A positive current means that the working electrode is behaving as the cathode.

- The oscillating current in Fig. 5.4 is due to the growth and fall of the Hg drops.
- As the drop grows, its area increases; more solute can reach the surface in a given time, and more current flows. The current increases as the drop grows until, finally, it falls off, and the current decreases sharply.

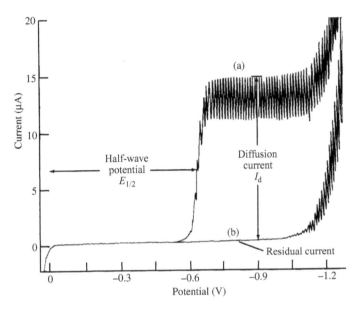

Figure 5.4 A graph of current versus potential in a polarogram.

5.4.1 Pulse Voltammetric Techniques

The primary advantage of pulse voltammetric techniques, such as normal pulse or differential pulse voltammetry, is their capacity to differentiate between capacitive current and faradaic current. Capacitive current (non-faradaic current) does not arise due to chemical reactions. It is produced due to the accumulation of electrical charges on the electrode and in the electrolyte solution near the electrode. Capacitive current is produced due to a change in the potential of an electrode [2]. Different types of waveforms used in pulse voltammetry are represented in Fig. 5.5.

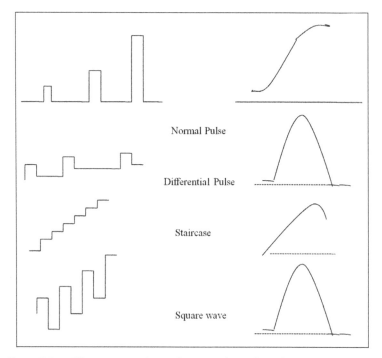

Figure 5.5 Different types of waveforms used in pulse voltammetry.

5.5 Square Wave Voltammetry

In the family of pulse voltammetric techniques, SWV is the most advanced and sophisticated technique [3, 4]. It is predominantly interesting, because of the fast scan rate, large amplitude, and the potential to differentiate against the capacitive current [5]. In addition, this technique facilitates concurrent analysis of both reduction and oxidation processes. The currents measured at the end of oxidation half-cycles give the oxidative (forward) current component, while the currents measured at reduction half-cycles give the reduction (backward) current component. The net current in SWV is obtained as a subtraction between the forward and the backward currents. However, reductive currents have a negative sign. The net current in SWV is actually a sum of the absolute values of both the current components. The net peak current is linearly proportional to the analyte concentration. It is a very rapid technique, providing insight into the kinetics of fast electron transfer reactions and into the kinetics of rapid chemical reactions coupled to the electroactive species. The advantage of SWV is that a response can be found at a high effective scan rate, thus reducing the scan time and lowering the consumption of electroactive compounds in relation to DPV. SWV shows wider linear range and a lower limit of detection because of its capability to discriminate capacitive current. SWV shows amplified current response in comparison to DPV response. SWV responses take only 1–5 s, while DPV responses take longer response times [6–10]. As SWV measurement is very fast, irreversible processes do not give a significant current signal [11]. This technique is suitable only for reversible systems. This feature can be of advantage when a sample contains two electroactive species in its matrix, which are reduced at similar potentials. If the electrode reaction of one of the species is reversible and that of the other irreversible, then the former can be readily detected in the presence of the latter, which gives no signal in SWV (Fig. 5.6).

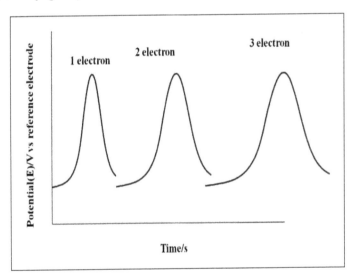

Figure 5.6 SWV curve for the number of electrons involved in the electrode reaction.

5.6 Differential Pulse Voltammetry

Among all techniques, DPV is most sensitive because in this approach, capacitive current is strongly differentiated with respect to faradaic current. It is a very sensitive technique since it can measure even a minute amount of electroactive species in the solution. The potential form in DPV consists of small pulses of constant amplitude (10–100 mV) superimposed on a staircase-wave form. The current in this technique is measured twice in each pulse period, first at a potential at the beginning of the applied pulse and second at the ending of the same pulse [12]. The measured current in the instrumental output, referred to as differential pulse voltammogram, is actually the difference between the currents measured for each single pulse. Irreversible systems have lower and broader peak currents with less sensitive and lower resolution than reversible systems [13–22]. For reversible reaction systems, the peak width ($W_{1/2}$) at half of peak height gives a value of 90.4 mV for one electron exchange ($W_{1/2} = 90.4/n$ mV), but for two-electron exchange, it is 45.2 mV (Fig. 5.7). The curves obtained between the changes in current and potential are peak shaped, and the height of the peak is directly proportional to the concentration of the analyte (Fig. 5.8). The DPV technique can be used for the simultaneous analysis of more than one electroactive species if the peak potentials are sufficiently far apart (normally at least 0.15 V). When the peak potentials differ by less than 0.15 V, the peaks overlap [23].

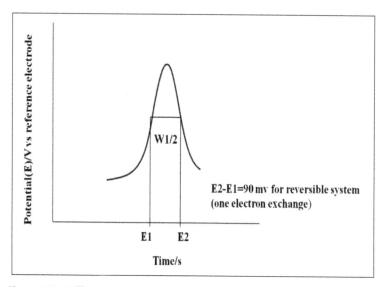

Figure 5.7 Differential pulse voltammogram.

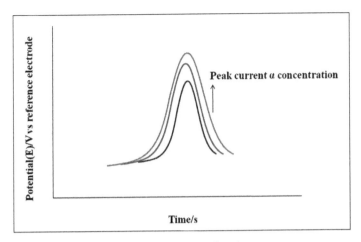

Figure 5.8 DPV of different concentrations of analyte.

This technique is very sensitive for the detection of trace amounts in drug analysis and forensic science. This technique is preferred to other techniques because of its distinct advantages.

5.7 Conclusion

Various electrochemical techniques are available, such as cyclic voltammetry, pulse voltammetry, SWV, and DPV, for the measurement of sensor response. Cyclic voltammetry is the most widely used since it measures current as a function of the applied potential and thus provides the mechanism of redox process. Pulse voltammetry is unique in its capacity to differentiate between capacitive and faradic currents. SWV is the most advanced technique because of the fast scan rate, large amplitude and potential to differentiate from the capacitive current. It also provides an analysis of both the reduction and oxidation processes simultaneously. DPV is one of the most sensitive techniques for the measurement of trace amounts of drugs and forensic materials.

Problems

1. What does the curve between the peak currents and the square root of the scan rates signify?

2. How to find out whether a process is because of electron transfer and diffusion is the means of mass transport?
3. What does the curve between log I_p versus log v signify?
4. What are the probabilities of noise in a cyclic voltammogram?
5. What is the significance of scan rate studies in any electrochemical study?
6. How can we calculate the surface area of the electrode using cyclic voltammetry?

References

1. Manea, F. (2014). Electrochemical techniques for characterization and detection application of nanostructured carbon composite. In: Aliofkhazraei, M. (ed.), *Modern Electrochemical Methods in Nano, Surface and Corrosion Science* (InTech, Rijeka, Croatia), http://dx.doi.org/10.5772/58633

2. Stojek, Z. (2001). Pulse voltammetry. In: Scholz, F. (ed.), *Electroanalytical Methods: Guide to Experiments and Applications* (Springer, Berlin Heidelberg, New York), pp. 99–110.

3. Lovric, M. (2001). Square-wave voltammetry. In: Scholz, F. (ed.), *Electroanalytical Methods: Guide to Experiments and Applications* (Springer, Berlin Heidelberg, New York), pp. 111–136.

4. Osteryoung, J.G., and O'Dea, J.J. (1986). Square-wave voltammetry. In: Bard, A.J. (ed.), *Electroanalytical Chemistry* (Marcel Dekker, New York).

5. Lavagnini, I., Antiochia, R., Magno F. (2004). An extended method for the practical evaluation of the standard rate constant from cyclic voltammogram, *Electroanalysis*, **16**, pp. 505.

6. Mirceski, V., Komorsky-Lovric, S., and Lovric, M. (2007). *Square Wave Voltammetry: Theory and Application* (Springer-Verlag, Berlin Heidelberg).

7. O'Dea, J.J., Osteryoung, J., and Osteryoung, R.A. (1981). Theory of square wave voltammetry for kinetic systems, *Anal. Chem.*, **53**, pp. 695.

8. Montenegro, M.I., Queiras, M.A., and Daschbach, J.L. (eds.) (1990). *Microelectrodes: Theory and Applications* (Kuwer Academic Publishers, the Netherlands).

9. Kalousek, M.A. (1948). Study of reversibility of processes at the dropping mercury electrode by changing discontinuously the polarizing voltage, *Collect. Czech. Chem. Commun.*, **13**, pp. 105.

10. Osteryoung, J.G., and Osteryoung, R.A. (1985). Square wave voltammetry, *Anal. Chem.*, **57**, pp. 101.

11. Stojek, Z., and Osteryoung, J.G. (1981). Direct determination of chelons at trace levels by square-wave polarography, *J. Anal. Chem.*, **53**, pp 847.

12. Osteryoung, R.A., and Osteryoung, J.G. (1981). *Phil. Trans. R. London.*, **A302**, 315–326.

13. The United States Pharmacopoeia, USP 23. (1994). *Validation of Compendial Methods. General Information* (The United States Pharmacopoeial Convention Inc., Taunton, Massachusetts).

14. Food and Drug Administration. (1987). *Guidelines for Submitting Samples and Validation Data for Methods Validation* (Food and Drug Administration, Rockville, Maryland). http://sfata.org/wp-content/uploads/2014/11/FDA_Guidances-Drugs_FINAL-_-Guidelines-for-Submitting-Samples-and-Analytical-Data-for-Methods-Validation.pdf

15. ICH Guideline Q2B. (1996). Validation of analytical procedures: Methodology.

16. Riley, C.M., and Rosanske, T.W. (eds.) (1996). *Development and Validation of Analytical Methods* (Pergamon, Oxford).

17. ICH Guideline Q2 (R1). (2005). Validation of analytical procedures: Text and methodology.

18. Lunte, S.M., and Radzik, D.M. (eds.) (1996). *Pharmaceutical and Biomedical Applications of Capillary Electrophoresis* (Pergamon, Oxford).

19. Haider, S.I. (ed.) (2002). *Pharmaceutical Master Validation Plan* (CRC Press, New York).

20. Thompson, M., Ellison, S.L.R., and Wood, R. (2002). Harmonized guidelines for single laboratory validation of methods of analysis, *Pure. Appl. Chem.*, **74**, pp. 835–855.

21. Berry, I.R., and Harpaz, D. (eds.) (2001). *Validation of Active Pharmaceutical Ingredients* (CRC Press, Boca Raton, Florida).

22. Bliesner, D.M. (ed.) (2006). *Validating Chromatographic Methods: A Practical Guide* (Wiley-Interscience, Hoboken, New Jersey).

23. Ozkan, S.A., Kauffmann, J.-M., and Zuman, P. (2015). Electroanalytical techniques most frequently used in drug analysis. In: *Electroanalysis in Biomedical and Pharmaceutical Sciences: Voltammetry, Amperometry, Biosensors, Applications* (Springer Berlin Heidelberg), pp. 45–81.

Chapter 6

Biosensors for Serum Metabolites

Jagriti Narang,[a] Nitesh Malhotra,[b] and Chandra Shekhar Pundir[c]

[a]*Amity Institute of Nanotechnology, Amity University, Noida-201313, India*
[b]*Amity Institute of Physiotherapy, Amity University, Noida-201313, India*
[c]*Department of Biochemistry, MD University, Rohtak-124001, India*
jags_biotech@yahoo.co.in

Electrochemical sensors for clinical analysis offer many advantages compared to a clinical laboratory setup, as these are fast, sensitive, specific, self-operable without training, and available at patients' bedside. Any serum metabolites in doses above or below the requisite quantity can result in serious adverse effects among individuals. Therefore, sensors are the best alternative for correct assessment and reducing chances of inappropriate treatment options.

Conventional methods used for the determination of serum metabolites involve expensive reagents and large sample volume. These methods have low specificity and sensitivity. On the other hand, electrochemical sensors offer high specificity and sensitivity.

This chapter discusses electrochemical sensors used for determining serum metabolites. The hub of this chapter lies on enzyme-based biosensors or electrochemical quantitative determination of

Biosensors: An Introductory Textbook
Edited by Jagriti Narang and Chandra Shekhar Pundir
Copyright © 2017 Pan Stanford Publishing Pte. Ltd.
ISBN 978-981-4745-94-9 (Hardcover), 978-1-315-15652-1 (eBook)
www.panstanford.com

serum metabolites. Biosensors for the determination of various se-rum metabolites can be classified on the basis of analytes and the mode of sensing [1–7].

Electrochemistry entails the transfer of charge from an electrode to another phase, which can be a solid or a liquid sample [8]. Upon the interaction of analytes and biological components immobilized on the electrode, chemical changes take place at the electrodes and are detected by the transducer (Fig. 6.1). The detector element works in a physicochemical way, like optical, piezoelectrical, electrochemical, or thermal, which transforms the signal resulting from the interaction of the analyte with the biological element into another signal that can be easily measured and quantified [9]. The biosensor can generate either distinct or continuous sensing signals, which are linearly related to the analyte content [10].

Different transducers have been applied in the fabrication of biosensors [11]. Biomolecules such as enzymes, DNA, antigen, antibody, and aptamers modified electrodes are a striking choice due to ease of fabrication, rapidity, high specificity, less pretreatment, and no requirement of skilled personnel. Electrochemical sensors monitor the current generated when electrons are exchanged either directly or indirectly between a biological system and an electrode [12].

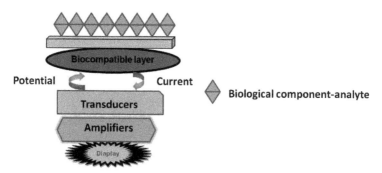

Figure 6.1 Schematic representation of biosensing device.

Biosensors can be categorized according to their transducing system. On the basis of the transducer assembly, biosensors are of the following types: (i) field effect based, (ii) optical, (iii) piezoelectric, (iv) calorimetric, and (v) electrochemical.

6.1 Biosensors Based on Field Effect

Field effect transistors (FETs) based on semiconductor devices are widely used in the fabrication of miniaturized chips. FETs respond to changes in the electric field. They can detect variations in the concentration of ions, as the gate region is exposed to an ionic solution [8]. Sensors in which there is a change in ionic concentration upon interaction of the analyte and the bioreceptor are most suited for FET-based biosensors, for instance urea biosensors in which there is a change in pH upon interaction (Fig. 6.2).

Advantages: These biosensors can be miniaturized, have high sensitivity and low cost, and possess multi-detection potential.

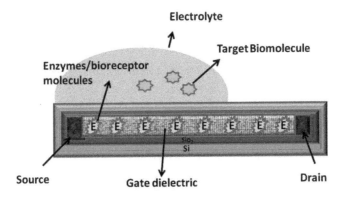

Figure 6.2   Schematic representation of FET-based biosensing device.

6.2 Optical Biosensors

In optical biosensors, fiber optics are utilized to measure the change upon interaction of analytes and bioreceptors such as proteins and nucleic acids. This change can be observed in both their physicochemical and optical properties. These biosensors measure the change in optical signals, which are linearly related to the analyte concentration (Fig. 6.3).

Advantages: These biosensors require no reference electrode, have a multianalyte detection system and insignificant interference, and can be miniaturized.

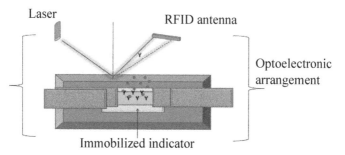

Figure 6.3 Schematic representation of an optical biosensing device.

6.3 Piezoelectric Biosensors

Piezoelectric biosensors are based on a mass sensitive detection system [11]. In these biosensors, transducers are utilized for the immobilization of bioreceptors on quartz crystals. These sensors are based on the principle of change in the resonant frequency due to the interaction between bioreceptors and analytes. This change is detected and converted into a measurable signal, for example resonance frequency changes.

Advantages: These are inexpensive and robust and have fast response.

6.4 Calorimetric Biosensors

In calorimetric biosensors, enzymes are combined with temperature sensors to measure the change upon interaction of analytes and bioreceptors. This change is observed in the form of amount of heat generated or consumed during a biochemical reaction. This biosensor measures the heat of reaction of the enzyme, which is linearly related with the analyte concentration.

6.5 Electrochemical Biosensors

Electrochemical biosensors have fascinated many biologists because of many distinct features such as high sensitivity, selectivity, and capability for miniaturize action. Biosensors detect analytes fast and cheap, with great specificity and sensitivity, when they get miniaturized.

Electrochemical biosensors are exploited in the determination of various serum metabolites such as bilirubin, glucose, ascorbic acid, cholesterol, triglycerides, oxalate, and urea. These biosensors are based on their specific bioreceptors, and change is produced after interaction of bioreceptors and analytes. This change is linearly proportional to the concentration of the analyte. The sensing signal is produced in the form of current, potential, and conductivity.

Potentiometric sensors generate a potential upon interaction of analytes and bioreceptors, which is proportional to the logarithm of the analyte concentration (Fig. 6.4). This potential is compared with the reference electrode potential at virtually zero current flow.

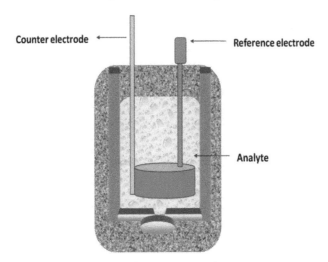

Figure 6.4 Schematic outline of potentiometric biosensors.

Conductometric biosensors entail either consumption or production of charged species and, therefore, lead to a global change in the ionic composition of the tested sample [11]. Thus, conductometric biosensors can detect any reactive change occurring in a solution. Formaldehyde, pesticides, insecticides, and nitrate biosensors are also developed conductometrically [11].

6.6 Applications of Biosensors for Determining Serum Metabolites

6.6.1 Glucose

Determining the concentration of blood glucose is very important for the clinical diagnosis of various diseases such as diabetes. Overdose of glucose can also induce the risk of renal, retinal, and neural complications, so the detection of glucose in human blood is medically important for the diagnosis of diabetes. Earlier methods used for determination have some setbacks such as pretreatment of sample, requirement of trained personnel to operate, lack of sensitivity and specificity, and long response time. Biosensors are the best alternative for the detection of blood glucose because of their distinct features such as high specificity, sensitivity, wide linear range, low response time, and enhanced operational and thermal stability. Enzyme-based biosensors offer sensitivity and specificity and consistent blood glucose monitoring in clinical samples [13–16]. The major drawbacks of enzyme-based sensors include improperly oriented enzyme immobilization, ambient conditions, short shelf life of enzymes, and high cost [17, 18]. To overcome these drawbacks, non-enzymatic sensors have been developed for better shelf life and selectivity toward the oxidation of glucose. Non-enzymatic sensors can be fabricated using various metal nanoparticles [19–25] and metal nanocomposites [25–34].

Sensors based on nanometal oxide provide many advantages in terms of sensitivity, specificity, and response time. Different morphologies of nanomaterials are available, which offer large surface area, promote electron transfer, and provide a compatible microenvironment to the biomolecules. Extensive efforts have been made to produce a reliable and robust method for the detection of glucose by nanometal oxides and their composites [35–37]. Various morphologies of zinc nanomaterials, such as nanowires [38], nanorods [39], nanocombs [40], and nanoclusters [41], have been exploited for the determination of glucose. All these zinc-based sensors are enzymatic sensors, because zinc does not show redox transition. Zinc-based glucose sensors provide better enzyme immobilization, high sensitivity, and better response time. Zinc oxide (ZnO) nano-materials have inherent advantages, but they also have some limita-tions, as these nanomaterials are easily detached from the sensing interface upon fictionalization [42, 43]. Extensive efforts have been made to overcome this drawback using ZnO nanocomposites, such as use of multi-walled carbon nanotubes (MWCNTs), as it provides better mechanical strength and conductivity [44, 45]. The use of ZnO-based nanocomposites provides better sensitivity, robustness, and conductivity. However, another major problem still persists with this nanomaterial; a relatively high potential is mandatory for opera-tion, which might involve oxidation of the interfering compounds. Therefore, at high potentials, sensors may exhibit some pseudo re-sults of the interfering species. Enzymatic sensors also have some disadvantages such as short shelf life and enzyme inactivity.

Researchers have used some more metal oxides, such as of copper and ruthenium, which show redox transition for sensing glucose. Copper-based nanomaterials offer direct oxidation of glucose without the use of enzymes and surface poisoning [46, 47]. Moreover, copper nanoparticles exhibit advantages such as promotion of faster electron transfer, catalytic properties, and low cost of production [48–50].

The exploitation of redox transition of copper oxide (CuO) in relation to non-enzymatic glucose oxidation leads to sensitive response and good stability [51]. Furthermore, CuO-based nanomaterials have been found to be highly stable and unaffected by the product of glucose oxidation. Different morphologies of copper have been exploited, such as CuO nanofibers [52], CuO nanowires [53], and flower-shaped CuO [54] for sensing glucose. Copper nanoparticles have one major limitation, oxygen sensitivity, which can make variations in results. Therefore, introduction of other nanomaterials in copper-based nanocompounds avoids interference from air. Researchers have used many copper-based nanocomposites for promoting the electron transfer rate of copper oxide. Introduction of MWCNT showed higher sensitivity and lower detection limit [53, 54].

Biosensors have one major advantage: these can be readily available near the patient's bedside. Biosensors should provide accurate results; otherwise, an inaccurate assessment of analytes can cause inappropriate options of treatment. Nowadays glucose levels are monitored in the home using disposable screen-printed enzyme electrode test strips [55, 56]. This technology involves printing of patterns of conductors and insulators on the surface of planar solid substrates. Each strip contains the printed working and reference electrodes, with the working one coated with the necessary reagents and membranes. This screen-printed enzyme electrode has attracted many researchers because of its high degree of superiority and low cost. Therefore, this modified electrode can be used for the fabrication of other sensors for determining various serum metabolites. But there is still a need for an ideal sensor, which can present a consistent real-time continuous monitoring of all blood glucose variations throughout the day with high selectivity [57]. The challenges for meeting these demands still persist and are open for researchers to produce highly stable biosensors, which can work under critical conditions.

6.6.2   Uric Acid

Uric acid is an important serum metabolite, and its determination is important for the diagnosis of diseases such as gout, Lesch–Nyhan syndrome, cardiovascular disease, uric acid kidney stones, and metabolic syndrome. Low concentration of urea causes multiple sclerosis (MS). Various techniques are available for the determination of uric acid in clinical samples, such as colorimetric, titrimetric, UV spectrophotometric, and mass spectrophotometric techniques. Since earlier methods have some setbacks, biosensing methods have been introduced for the determination of uric acid to overcome the limitations. At present, biosensing techniques are most often used because these are fast and highly specific.

Various artificial membranes and nanomaterials have been exploited for the electroanalysis of uric acid, such as metal nanoparticles and nanocomposites. Artificial membranes used for the electrodetermination of uric acid are egg-shell membrane [58] and silk fibroin membrane [59]. Membrane-based biosensors offer a portable, economical, and fast method. Distinct features make them capable matrices for enzyme immobilization, which can enhance the sensitivity and selectivity of biosensors [60]. Major disadvantages of artificial membranes are fragility, leakage of enzyme, and adherence of enzymes to the electrode. To overcome these setbacks, researchers have employed nanomaterials and nanocomposites for the determination of uric acid. Nanostructured metal oxides have some distinct features such as high surface area as well as faster electron transfer kinetics between the electrode and the active site of the desired enzyme [61–66]. Nanomaterials-modified interfaces have distinct advantages, for instance these are highly sensitive and specific, provide a biocompatible environment, are less prone to surface fouling, and show redox activity at a lower potential, which eliminates the interfering species [67].

Various polymers are used for the electroanalysis of uric acid, such as polyaniline [68–70], polypyrrole [71], poly(o-aminophenol) [72], and pani-polymethacrylate [73]. Polymers are conducive, provide high surface-to-volume ratio and efficient transfer of electrons, and are suitable for biosensing applications. There is one limitation in using polymers: low robustness. To increase the mechanical strength of polymers, nanoparticles are introduced in the polymer matrix. Furthermore, nanomaterials of various morphologies induce and promote conductivity and mechanical strength, and high surface-to-volume ratio increases the loading amount of biomolecules. Nanomaterials used for the electroanalysis of uric acid are MWCNT with SnO_2 nanoparticles [74], ZnO nanorods [75], ZnS quantum dots [76], and ruthenium/oxidepyrochlor [77].

Nowadays screen-printed electrode technology has provided a whirling point for the electrochemical determination of uric acid [78, 79]. Enzyme-less determination of uric acid offers many distinct advantages such as increased shelf life, low cost of enzyme, and no effect of variable conditions such as temperature and pH [80]. Determination and management of uric acid are increasingly important for the point of care and patients' self-monitoring. Currently, the majority of the portable uric acid monitoring devices on the market are mostly based on electrochemically modified technology. According to the explanations of the currently available electrochemical uric acid monitoring systems, almost all apply the non-enzymatic method [81].

6.6.3 Glycated Hemoglobin

Diabetes mellitus is a chronic metabolic disorder characterized by a rise in the blood glucose level, called hyperglycemia [82].

Determination of the whole glucose level is essential for patients' self-monitoring to maintain it within a normal range. Overconcentration of this serum metabolites causes serious ailments such as kidney failure, blindness, and nerve damage [83]. The glucose present in blood reflects a fluctuating glucose level, and it is affected by daily diet and requires frequent measurements. Therefore, an accurate analysis of blood glucose level is necessary so that patients can adopt appropriate options of treatment or precautions at the earliest. Determination of the blood glycated hemoglobin (HbA1C) level is necessary for the accurate and early diagnosis of diabetes mellitus. The compound level is detected by measuring the ratio between the HbA1C concentration and the total Hb concentration [84]. The HbA1C level acts as an early diagnostic index, unaffected by fluctuation, and allows selecting an appropriate option of treatment. Therefore, the amount of blood HbA1C level is important for the lasting management of the glycemic state in diabetic patients [85].

Various techniques are now available for the analysis of HbA1C level, such as high-performance liquid chromatography (HPLC) [86–88], ion exchange chromatography [89], liquid chromatography associated in tandem with mass spectroscopy [85], immunoassay [90], and colorimetric [91]. Earlier methods offer many advantageous features such as high sensitivity, reproducibility, and accuracy. But in HPLC, defined peak assigned for HbA1C is not scientifically acceptable, as it lacks specificity. Furthermore, earlier methods require advanced instrumentation and skilled person, and understanding the results of the peaks obtained is comparatively difficult. Biosensors can overcome all disadvantages associated with the earlier methods. Researchers have employed various membranes for determining the HbA1C level, such as CDI-activated cellulose membranes [92], membrane-modified Pt electrode [93], and polyvinyl alcohol (PVA) [94]. Conducting polymers and nanoparticles were also used for the electroanalysis of glycated hemoglobin, such as polyvinylimidazole polymer [95], magnetic nanoparticles [96], iridium-modified carbon electrode [97], zirconium dioxide (ZrO_2) nanoparticles [98], ZnO nanoparticles [99], and graphene oxide [100].

All aforementioned biosensors are based on enzymes, i.e., fructosyl valine oxidase. Enzymatic biosensors have some limitations such as less stability, difficulty in fabrication, and high price. To overcome these limitations, non-enzymatic sensors are also employed for the determination of HbA1C. The enzyme-less method for the determination of HbA1C is advantageous in terms of stability, low oxidation potential, low price, and ease of fabrication of the materials involved. In conclusion, the non-enzymatic method can generate an amplified signal with low potential and high stability. These types of biosensors are amenable to miniaturization and have compatible instrumental sensitivity [99]. Potentiometric sensors are also fabricated for the determination of HbA1C, for instance integrated chip and micro-extended gate electrode array [101], mixed self-assembled monolayers (SAMs) wrapped nanospheres array, FET sensor chip [102], and a disposable micro-electrode array chip [103].

Development of biosensors has provided many advantageous features compared to earlier detection methods and has played an important role in diabetes diagnosis and management. Miniaturization of a biosensor not only reduces the size of the detection device and sample volume, but also integrates all steps of the analytical process into a single-sensor device. Thus, it results in a reduction in both time and cost of analysis. Moreover, it is expected to lead to further portability for in vivo sensing and in field applications. Trials can be made to integrate the laboratory model with chip-designing companies (VLSI) in order to develop a fully automated chip system in a device, to be used by any patient on his or her bedside.

6.6.4 Triglycerides

Triglyceride (triacylglycerol, TAG, or triacylglyceride, TG) is an ester of three molecules of fatty acids and one mole of glycerol. A fatty acid consists of a long chain of hydrocarbon ending with a carboxyl group (−COOH). Triglycerides, as major components of very-low-density lipoprotein, play an important role in metabolism as energy source and act as transporters of dietary fat. Determination of triglycerides is important for the diagnosis and management of various diseases such as atherosclerosis, risk of heart disease, and stroke. The existing techniques available for the determination of triglyceride are chemical [104], HPLC [105, 106], mass spectrometry [107], microtiter plate method [108], and enzymic colorimetric and fluorimetric methods [109, 110], and enzyme electrodes are available [111–113]. Earlier methods, specifically HPLC, are highly specific and suitable for many applications but do not satisfy the necessities for a simple, fast, accurate, and specific analysis, as these are complicated and require time-consuming sample pretreatment, expensive instrumental setup, and skilled persons to operate [114].

Multi-enzymatic biosensors are available for the determination of TG. The multi-enzymes involved in the TG biosensing system are lipase, glycerol kinase, and glycerol-3-phosphate oxidase (GPO) [115–122]. Multi-enzymatic systems have limitations such as difficult co-immobilization procedures and requirement of many cofactors for specific enzymes. These biosensors can be sometimes expensive and time-consuming. Hence, TG biosensors are fabricated based on bienzymic (lipase and glyceroldehydrogenase) and one enzymatic (lipase) TG sensing.

In the fabrication of TG biosensors, various membranes are employed, such as egg-shell membrane [116], polyvinyl chloride [117], polyvinyl alcohol [119], collagen [123], autophagic vacuole membrane [124], and cellulose acetate [125]. Membranes-based biosensors are comparatively more simple, sensitive, and rapid. In multi-enzymatic TG sensors, membranes provide the best co-immobilization matrix for enzymes. But membranes have problems such as time-consuming preparation, long response time, fragility, and low storage stability. Above all, they have a major problem of desorption or leaching of enzymes during washing. Various conducting polymers are employed for the determination of TG, for instance polyaniline/single-walled carbon nanotubes [126].

Polymers have some distinct features such as biocompatibility, and they protect the sensing material from ambient conditions. Furthermore, they can electro-polymerize themselves on any surface of electrode. Researchers have used nanoparticles for better results in terms of conductivity, response time, sensitivity, and specificity. Nanoparticles are known to have the unique ability to promote faster electron transfer kinetics between the electrode and the active site of the desired enzyme [127–130]. Nanoparticles-modified electrodes have fascinated researchers, because of advantages such as high sensitivity, selectivity and stability, less prone to surface fouling, and low over-potential at which the electron transfer process occurs, compared with inert substrate electrodes

[131]. Various nanoparticles are employed, such as cerium oxide (CeO_2) [132], iridium nanoparticles [133], MWCNTs [134], gold polypyrrole nanocomposite [115], and ZnO nanoparticles [118]. Nanoparticles-based TG biosensors have many advantages both in terms of stability and in promoting the catalytic reduction of redox species. Additionally, the electrode is notable for its ability to inhibit the oxidation of interfering species. Nanoparticles have fascinated much attention due to their exceptional features such as high mechanical strength, oxygen ion conductivity, biocompatibility, increased electroactive surface, and retention of biological activities [135, 136].

Researchers have conducted broad studies on the electroanalysis of metabolites by observing the change in the charge density using ion-selective FETs (ISFETs) [137]. Different potentiometric sensors have also been developed for the determination of TG, such as mesoporus Si matrix [138], pH FET [139], porous silicon [140, 141], silica gel beads [142], gate surface of FET [143], and silicon nitride layer of electrolyte insulator semiconductor capacitors (EISCAP) and cantilever beams [144]. In this regard, an ISFET device of small size and low weight might be appropriate for use in a portable monitoring system, i.e., a hand-held metabolite-monitoring system.

6.6.5 Creatinine

Creatinine is a heterocyclic compound, which sustains the wellness of skeletal muscles and helps in supplying energy to them [145]. The determination of creatinine is essential, since the creatinine level in urine reveals the condition of renal and muscle function. Creatinine levels are used to assess kidney function as well as to detect other urinary tract disorders. High concentrations of creatinine in the serum indicate inadequate function of the kidneys, also known as acute/chronic renal failure. Creatinine is more important serum metabolites for the determination of kidney function than urea.

Different techniques are available for the determination of creatinine, which include enzymic colorimetric methods [146], HPLC [147], liquid chromatography–mass spectrometry [148], gas chromatography–mass spectrometry [149], and optical methods [150, 151]. A majority of these methods are accurate and appropriate for many applications. Conversely, these techniques do not satisfy the necessities for a rapid, precise, and specific analysis as these are complex, lengthy, and requires pretreatment of samples. Biosensors are the only solution to overcome these problems as they are fast, specific, and sensitive.

Membrane-based sensors have been reported for the determination of creatinine silicone gas-permeable membrane [152], cellulose acetate [153], gas-permeable membrane [154], and polyvinyl alcohol [155]. Membrane-based creatinine biosensors are easy to fabricate, readily available at the patient's bedside, and can be employed without training.

Various polymers are also employed for the determination of creatinine, for instance poly(γ-methyl-L-glutamate) [156], polypropylene [157], propylamine [158–160], polypyrrole [161], poly-2-hydroxyethyl methyacrylate [162], poly(1,3-diaminobenzene) [163], poly(carbamoyl) sulfonate-hydrogel matrix [164], polyaniline-nafion [165], and poly(carbamoyl) sulfonate-hydrogel matrix [166]. These polymers-based biosensors achieve exceptional operational stability, long storage lifetimes, relatively short response time, and high sensitivity.

Nanobiosensors have distinct features, for instance improved detection, sensitivity, and specificity, and possess great potential in applications on clinical samples [167, 168]. Nanomaterials employed for the detection of creatinine are ZnO nanoparticles and c-MWCNT [167] and Fe_3O_4 nanoparticles [168]. By incorporating carbon nanotubes, biosensors provide large surface area and promote electric properties and electrocatalysis. Metal oxide nanoparticles modified electrodes show distinct features such as better catalysis, amplification of detection signal, and large effective surface area.

Potentiometric sensors are also available for the determination of creatinine, such as carboxylated-polyvinyl chloride [169], CHIT-SiO_2-MWCNTs [170], CHIT-g-PANI [171], and chemical current conveyors [172]. Potentiometric sensors have the advantage of relative simplicity and are based on well-established gas-sensing electrode technologies; they also avoid interferences [173].

Non-enzymatic sensors are also available for the determination of creatinine. Non-enzymatic sensors offer some distinct advantageous features; they can be applied for regular urine analysis even in a small and simply equipped laboratory and provide large stability.

Researchers have developed non-enzymatic sensors, for instance copper–platinum electrode [174] and β-cyclodextrin incorporated poly-3,4-ethylene dioxythiophene modified glassy carbon electrode [175]. These sensors permit either identification of various urine samples or detection of several parameters of the same sample at the same time.

6.7 Conclusion

The initial impetus for advancing sensor technology came from the health care sector, where it is now generally recognized that measurements of blood gases, ions, and metabolites are often essential for a better estimation of the metabolic state of patients. Nowadays, there is need for regular monitoring of serum metabolites for appropriate management of diseases. Instant analysis of serum metabolites can provide early information of diseases. The relative negative impacts of raised levels of serum metabolites are linked to various chronic diseases. Biosensors provide alternative means for instant analysis at the patient's bedside. Electrochemical biosensors can possibly meet such needs provided their capabilities can be extended to a wider range of biomolecules and more complex matrices. Biosensors have the potential of a billion-dollar market, and the technology needs improvement in biological stability, signal transduction, precision, and cost effectiveness. In this chapter, attempts have been made to summarize the salient features of sensors based on various serum metabolites.

Problems

1. What is the significance of plotting the calibration curve of any analyte?
2. How to calculate the value of unknown concentration of an analyte?
3. What is the significance of the scan-rate studies in any electrochemical study?
4. How to calculate the analytical recovery in any electrochemical study?
5. How to calculate precision in any electrochemical study?

6. How to calculate accuracy in any electrochemical study?
7. What is the significance of performing the interference study?
8. How to find out the detection limit in any electrochemical study?
9. What do you mean by the specificity of an analytical system?

References

1. Daniel, S., Rao, T.P., Rao, K.S., Rani, S.U., Naidu, G.R.K., Lee, H.-Y., and Kawai, T. (2007). A review of DNA functionalized/grafted carbon nanotubes and their characterization. *Sens. Actuat. B Chem.,* **122**: 672–682.

2. Drummond, T.G., Hill, M.G., and Barton, J.K. (2003). Electrochemical DNA sensors. *Nat. Biotechnol.,* **21**: 1192–1199.

3. Dorazio, P. (2003). Biosensors in clinical chemistry. *Clin. Chim. Acta,* **334**: 41–69.

4. Marquette, C.A., and Blum, L.J. (2006). State of the art and recent advances in immunoanalytical systems. *Biosens. Bioelectron.,* **21**: 1424–1433.

5. Stefan, R.I., Staden, J.F., and Aboul-Enein, H.Y. (2000). Immunosensors in clinical analysis. *Fresenius J. Anal. Chem.,* **366**: 659–668.

6. Rivas, G.A., Rubianes, M.D., Rodriguez, M.C., Ferreyra, N.F., Luque, G.L., Pedano, M.L., Miscoria, S.A., and Parrado, C. (2007). Carbon nanotubes for electrochemical biosensing. *Talanta,* **74**: 291–307.

7. Vestergaard, M., Kerman, K., and Tamiya, E. (2007). An overview of label-free electrochemical protein. *Sensors,* **7**: 3442–3458.

8. Wang, Y., Xu, H., Zhang, J., and Li, G. (2008). Electrochemical sensors for clinic analysis. *Sensors,* **8**: 2043–2081.

9. Lei, Y., Chen, W., and Mulchandani, A. (2006). Microbial biosensors. *Anal. Chim. Acta,* **568**: 200–210.

10. Turner, A.P., Chen, B., and Piletsky, S.A. (1999). In vitro diagnostics in diabetes: Meeting the challenge. *Clin. Chem.,* **45**: 1596–1601.

11. Turner, A.P., Karube, I., and Wilson, G.S. (1987). *Biosensors: Fundamentals and Applications,* Oxford University Press, Oxford, 770.

12. Grieshaber, D., MacKenzie, R., Vörös, J., and Reimhult, E. (2008). Electrochemical biosensors—Sensor principles and architectures, *Sensors (Basel),* **8**: 1400–1458.

13. Koschinsky, T., and Heinemann, L. (2001). Sensors for glucose monitoring: Technical and clinical aspects. *Diabetes Metab. Res. Rev.,* **17**: 113–123.

14. Oliver, N.S., Toumazou, C., Cass, A.E.G., and Johnston, D.G. (2009). Glucose sensors: A review of current and emerging technology. *Diabetic Med.,* **26**: 197–210.

15. Lin, Y., Lu, F., Tu, Y., and Ren, Z. (2004). Glucose biosensors based on carbon nanotube nanoelectrode ensembles. *Nano Lett.,* **4**: 191–195.

16. Rakow, N.A., and Suslick, K.S. (2000). A colorimetric sensor array for odour visualization. *Nature,* **406**: 710–712.

17. Reitz, E., Jia, W., Gentile, M., Wang, Y., and Lei, Y. (2008). CuO nanospheres based nonenzymatic glucose sensor. *Electroanalysis,* **20**: 2482–2486.

18. Sachedina, N., and Pickup, J.C. (2003). Performance assessment of the Medtronic-MiniMed Continuous Glucose Monitoring System and its use for measurement of glycaemic control in Type 1 diabetes. *Diabet. Med.,* **20**: 1012–1015.

19. Proenca, L., Lopes, M.I.S., Fonseca, I., Kokoh, K.B., Leger, J.M., and Lamy, C. (1997). Electrocatalytic oxidation of D-sorbitol on platinum in acid medium: Analysis of the reaction products. *J. Electroanal. Chem.,* **432**: 237–242.

20. Chen, C.Y., Tamiya, E., Ishihara, K., Kosugi, Y., Su, Y.C., Nakabayashi, N., Karube, I.A. (1992). Biocompatible needle-type glucose sensor based on platinum-electroplated carbon electrode. *Appl. Biochem. Biotech.,* **36**: 211–226.

21. Chou, C.H., Chen, J.C., Tai, C.C., Sun, I.W., and Zen, J.M. (2008). A nonenzymatic glucose sensor using nanoporous platinum electrodes prepared by electrochemical alloying/dealloying in a water-insensitive zinc chloride-1-ethyl-3 methylimidazolium chloride ionic liquid. *Electroanalysis,* **20**: 771–775.

22. Kurniawan, F., Tsakova, V., and Mirsky, V.M. (2006). Gold nanoparticles in nonenzymatic electrochemical detection of sugars. *Electroanalysis,* **18**: 1937–1942.

23. Meng, L., Jin, J., Yang, G., Lu, T., Zhang, H., Cai, C. (2009). Nonenzymatic electrochemical detection of glucose based on palladium-single-walled carbon nanotube hybrid nanostructures. *Anal. Chem.,* **81**: 7271–7280.

24. Zhu, H., Lu, X., Li, M., Shao, Y., and Zhu, Z. (2009). Nonenzymatic glucose voltammetric sensor based on gold nanoparticles/carbon nanotubes/ionic liquid nanocomposites. *Talanta,* **79**: 1446–1453.

25. Wang, H., Zhou, C., Liang, J., Yu, H., Peng, F., and Yang, J. (2008). High sensitivity glucose biosensor based on Pt electrodeposition onto low-density aligned carbon nanotubes. *Int. J. Electrochem, Sci.*, **3**: 1258–1267.

26. Xie, J., Wang, S., Aryasomayajula, L, and Varadan, V.K. (2007). Platinum decorated carbon nanotubes for highly sensitive amperometric glucose sensing. *Nanotechnology*, **18**: 065503–065512.

27. Wang, H., Zhou, C., Liang, J., Yu, H., and Peng, F. (2008). An enhanced glucose biosensor modified by Pt/sulfonated-MWCNTs with layer by layer technique. *Int. J. Electrochem. Sci.*, **3**: 1180–1185.

28. Feng, D., Wang, F., and Chen, Z. (2009). Electrochemical glucose sensor based on one-step construction of gold nanoparticle–chitosan composite film. *Sens. Actuat. B Chem.*, **138**: 539–544.

29. Belousov, V.M., Vasylyev, M.A., Lyashenko, L.V., Vilkova, N.Y., and Nieuwenhuys, B.E. (2003). The low-temperature reduction of Pd-doped transition metal oxide surfaces with hydrogen. *J. Chem. Eng.*, **91**: 143–150.

30. Tominaga, M., Nagashima, M., Nishiyama, K., and Taniguchi, I. (2007). Surface poisoning during electrocatalytic monosaccharide oxidation reactions at gold electrodes in alkaline medium. *Electrochem. Commun.*, **9**: 1892–1898.

31. Liua, Y., Teng, H., Hou, H., and You, T. (2009). Nonenzymatic glucose sensor based on renewable electrospun Ni nanoparticle-loaded carbon nanofiber paste electrode. *Biosens. Bioelectron.*, **24**: 3329–3334.

32. Tian, K., Prestgard, M., Tiwari, A., (2014). A review of recent advances in nonenzymatic glucose sensors. *Mat. Sci. Eng. C*, **41**: 100–118.

33. Lu, L.M., Zhang, L., Qu, F.L., Lu, H.X., Zhang, X.B., Wu, Z.S., Huana, S.Y., Wang, Q.A., Shen, G.L., and Yu, R.Q. (2009). A nano-Ni based ultrasensitive nonenzymatic electrochemical sensor for glucose: Enhancing sensitivity through a nanowire array strategy. *Biosens. Bioelectron.*, **25**: 218–223.

34. Grace, A.N., and Pandian, K. (2007). Synthesis of gold and platinum nanoparticles using tetraaniline as reducing and phase transfer agent: A brief study and their role in the electrocatalytic oxidation of glucose. *J. Phys. Chem. Sol,* **68**: 2278–2285.

35. Holt-Hindle, P., Nigro, S., Asmussen, M., and Chen, A. (2008). Amperometric glucose sensor based on platinum–iridium nanomaterials. *Electrochem. Commun.,* **10**: 1438–1441.

36. Xiao, F., Zhao, F., Mei, D., Mo, Z., and Zeng, B. (2009). Nonenzymatic glucose sensor based on ultrasonic-electrodeposition of bimetallic PtM (M = Ru, Pd and Au) nanoparticles on carbon nanotubes–ionic liquid composite film. *Biosens. Bioelectron.*, **24**: 3481–3486.

37. Cui, H.F., Ye, J.S., Liu, X., Zhang, W.D., and Sheu, F.S. (2006). Pt–Pb alloy nanoparticle/carbon nanotube nanocomposite: A strong electrocatalyst for glucose oxidation. *Nanotechnology*, **17**: 2334–2339.

38. Zang, J., Li, C.M., Cui, X., Wang, J., Sun, X., Chang, H.D., and Sun, Q. (2007). Tailoring zinc oxide nanowires for high performance amperometric glucose sensor. *Electroanalysis*, **19**: 1008–1014.

39. Liu, X.W., Hu, Q., Wu, Q., Zhang, W., Fang, Z., and Xie, Q. (2009). Aligned ZnO nanorods: A useful film to fabricate amperometric glucose biosensor. *Colloid Surf. B Biointerfaces*, **74**: 154–158.

40. Wang, J.X., Sun, X.W., Wei, A., Lei, Y., Cai, X.P., Li, C.M., and Dong, Z.L. (2006). Zinc oxide nanocomb biosensor for glucose detection. *Appl. Phys. Lett.*, **88**: 233106(1–3).

41. Zhao, Z.W., Chen, X.J., Tay, B.K., Chen, J.S., Han, Z.J., and Khor, K.A. (2007). A novel amperometric biosensor based on ZnO: Co nanoclusters for biosensing glucose. *Biosens. Bioelectron.*, **23**: 135–139.

42. Basu, S., Kang, W.P., Davidson, J.L., Choi, B.K., Bonds, A.B., and Cliffel, D.E. (2006). Electrochemical sensing using nanodiamond microprobe. *Diam. Relat. Mat.*, **15**: 269–274.

43. Rodriguez, J.A., Jirsak, T., Dvorak, J., Sambasivan, S., and Fischer, D. (2000). Reaction of NO_2 with Zn and ZnO: Photoemission, XANES, and density functional studies on the formation of NO_3. *J. Phys. Chem. B*, **104**: 319–328.

44. Farazmand P.M., Khanlary, S., Fehli, A., and Elahi, S.M. (2015). Synthesis of carbon nanotube and zinc oxide (CNT–ZnO) nanocomposite. *J. Inorgan. Organometall. Poly. Mat*, **4**: 942–947.

45. Alharbi, N.D., Ansari, M.S., Salah, N., Khayyat, S.A., and Khan, Z.H. (2016). Zinc oxide-multi walled carbon nanotubes nanocomposites for carbon monoxide gas sensor application. *J Nanosci. Nanotechnol.*, **16**: 439–447.

46. Mho, S.I., and Johnson, D.C. (2001). Electrocatalytic response of carbohydrates at copper-alloy electrodes. *J. Electroanal. Chem.*, **500**: 524–532.

47. Salimi, A., and Roushani, M. (2005). Non-enzymatic glucose detection free of ascorbic acid interference using nickel powder and nafion sol–gel dispersed renewable carbon ceramic electrode. *Electrochem. Commun.*, **7**: 879–887.

48. Gao, X.P., Bao, J.L., Pan, G.L., Zhu, H.Y., Huang, P.X., Wu, F., and Song, D.Y. (2004). Preparation and electrochemical performance of polycrystalline and single crystalline CuO nanorods as anode materials for Li ion battery. *J. Phys. Chem. B,* **108**: 5547–5551.

49. Zhang, J., Liu, J., Peng, Q., Wang, X., and Li, Y. (2006). Nearly monodisperse Cu_2O and CuO nanospheres: Preparation and applications for sensitive gas sensors. *Chem. Mater.,* **18**: 867–871.

50. Rakhshani, A.E., Makdisi, Y., and Mathew, X. (1996). Deep energy levels and photoelectrical properties of thin cuprous oxide films. *Thin Solid Films,* **288**: 69–75.

51. McAuley, C.B, Du, Y., Wildgoose, G.G., and Compton, R.G. (2008). The use of copper(II) oxide nanorod bundles for the non-enzymatic voltammetric sensing of carbohydrates and hydrogen peroxide. *Sens. Actuat. B Chem.,* **135**: 230–235.

52. Wang, W., Zhang, L., Tong, S., Li, X., and Song, W. (2009). Three-dimensional network films of electrospun copper oxide nanofibers for glucose determination. *Biosens. Bioelectron.,* **25**: 708–714.

53. Zhuang, Z., Su, X., Yuan, H., Sun, Q., Xiao, D., and Choi, M.M.F. (2008). An improved sensitivity non-enzymatic glucose sensor based on a CuO nanowire modified Cu electrode. *Analyst,* **133**: 126–132.

54. Jiang, L.C., and Zhang, W.D. (2010). A highly sensitive nonenzymatic glucose sensor based on CuO nanoparticles-modified carbon nanotube electrode. *Biosens. Bioelectron.,* **25**: 1402–1407.

55. Xu, H., Li, G., Wu, J., Wang, Y., and Liu, J. (2005). A glucose oxidase sensor based on screen-printed carbon electrodes modified by polypyrrole. *J Conf Proc IEEE Eng Med Biol Soc.,* **2**: 1917–1920.

56. Newman, J.D., and Turner, A.P. (2005). Home blood glucose biosensors: A commercial perspective. *Biosens. Bioelectron.,* **20**: 2435–2453.

57. Wang, J. (2008). Electrochemical glucose biosensors. *Chem. Rev.,* **108**: 814–825.

58. Zhang, Y., Wen, G., Zhou, Y., Shuang, S., Dong, C., and Choi, M.M. (2007). Development and analytical application of a uric acid biosensor using an uricase-immobilized eggshell membrane. *Biosens. Bioelectron.,* **22**: 1791–1797.

59. Zhang, Y.Q., Shen, W.D., Gu, R.A., Zhu, J., and Xue R.Y. (1998). Amperometric biosensor for uric acid based on uricase immobilized silk fibrion membrane. *Anal. Chim. Acta,* **369**: 123–128.

60. Pundir, C.S., and Narang, J. (2013). Determination of triglycerides with special emphasis on biosensors: A review. *Int. J. Biol. Macromol.,* **61**: 379–389.

61. Pandey, P., Datta, M., and Malhotra, B.D. (2008). Prospects of nanomaterials in biosensor. *Anal. Lett.*, **41**: 157–207.

62. Zhou, H., Gan, X., Wang, J., Zhu, X., and Li, G. (2005). Hemoglobin-based hydrogen peroxide biosensor tuned by the photovoltaic effect of nano titanium dioxide. *Anal. Chem.*, **77**: 6102–6104.

63. Kumar, S.A., and Chen, S.M. (2008). Nanostructured zinc oxide particles in chemically modified electrodes for biosensor applications. *Anal. Lett.*, **41**: 141.

64. Singh, S.P., Arya, S.K., Pandey, M.K., Malhotra, B.D., Saha, S., Sreenivas, K., and Gupta, V. (2007). Cholesterol biosensor based rf sputtered zinc oxide nanoporous thin films. *Appl. Phys. Lett.*, **91**: 063901.

65. Anees, A., Ansari, M., Alhoshan, M.S., and Alsalhi, A.S. (2010). Nanostructured metal oxides based enzymatic electrochemical biosensors biomedical engineering, In: *Biosensors*, Pier Andrea Serra (Ed).

66. Wei, A., Sun, X.W., Wang, J.X., Lei, Y., Cai, X.P., Li, C.M., Dong, Z.L., and Huang, W. (2006). Enzymatic glucose biosensor based on ZnO nanorod array grown by hydrothermal decomposition. *Appl. Phys. Lett.*, **99**: 123902.

67. Premamoy, G., Samir, K.S., and Amit, C. (1999). Characterization of poly(vinyl pyrrolidone) modified polyaniline prepared in stable aqueous medium. *Eur. Polym. J.*, **35**: 699–710.

68. Jiang, Y., Wang, A., and Kan, J. (2007). Selective uricase biosensor based on polyaniline synthesized in ionic liquid. *Sens. Actuat. B Chem.*, **124**: 529–534.

69. Arora, K., Sumana, G., Saxena, V., Gupta, R.K., Gupta, S.K., Yakhmi, J.V., Pandey, M.K, Chand, S., and Malhotra, B.D. (2007). Improved performance of polyaniline-uricasebiosensor. *Anal. Chim. Acta*, **594**: 17–23.

70. Kan, J., Pan, X., and Chen, C. (2004). Polyaniline-uricase biosensor prepared with template process. *Biosens. Bioelectron.*, **19**: 1635–1640.

71. Cete, S., Yasar, A., and Arslan, F. (2006). An amperometric biosensor for uric acid determination prepared from uricase immobilized in polypyrrole film. *Artif. Cells Blood Substit. Immobil. Biotechnol.*, **34**: 367–380.

72. Miland, E., Miranda Ordieres, A.J., Tunon, B.P., Smyth, M.R., and Fagain, C.O. (1996). Poly(o-aminophenol)-modified bienzyme carbon paste electrode for the detection of uric acid. *Talanta*, **43**: 785–796.

73. Castillo-Ortega, M.M., Rodriguez, D.E., Encinas, J.C., Plascencia, M., Mendz, F.A., and Olayo, R. (2002). Conductometric uric acid and urea

biosensor prepared from electroconductive polyaniline poly(n-butyl methacrylate) composites. *Sens. Actuat. B Chem.,* **85**: 19–25.

74. Zhang, F.F., Wang, X.L., Li, C.X., Li, X.H., Wan, Q., and Xian, Y.Z. (2005). Assay for uric acid level in rat striatum by a reagent-less biosensor based on functionalized multi-wall carbon nanotubes with tin oxide. *Anal. Bioanal. Chem.,* **382**: 1368-1373.

75. Zhang, F., Wang, X., Shiyun A.I., Sun, Z., Wan, Q., Zlu, Z., Xian, Y., and Jin, L. (2004). Immobilization of uricase on ZnO nanorods for a reagent-less uric acid sensor. *Anal. Chim. Acta,* **519**: 155–160.

76. Zhang, F., Li, C., Li, X., Wang, X., Wan, Q., and Xian, Y. (2006). ZnS quantum dots derived a reagent-less uric acid biosensor. *Talanta,* **68**: 1353–1358.

77. Zen, J.M., Lai, Y.Y., Yang, H.H., and Senthil, K.A. (2002). Multianalyte sensor for the simultaneous determination of hypoxanthine, xanthine and uric acid based on perionized nontronite coated screen printed electrode. *Sens. Actuat. B Chem.,* **84**: 237–244.

78. Gilmartin, M., Hart, J., and Birch, B. (1994). Development of amperometric sensors for uric acid based on chemically modified graphite-epoxy resin and screen-printed electrodes containing cobalt phthalocyanine. *Analyst,* **119**: 243–252.

79. Gilmartin, M., and Hart, J. (1995). Development of one-shot biosensors for the measurement of uric acid and cholesterol. *Anal. Proc. Incl. Anal. Commun.,* **32**: 341–345.

80. Zen, J., Jou, J., and Ilangovan, G. (1998). Selective voltammetric method for uric acid detection using pre-anodized nafion-coated glassy carbon electrodes. *Analyst,* **123**: 1345–1350.

81. Ting, L., Chi, C., Liao, C., Ching, L., Ting, Y., Yang, G., and Chueng, W. (2014). Evaluation of an electrochemical biosensor for uric acid measurement in human whole blood samples. *Clinica. Chimica. Acta,* **436**: 72–77.

82. Vinik, A., and Flemmer, M. (2001). Diabetes and macrovascular disease. *J. Diabetes Complicat.,* **16**: 235–245.

83. Genuth, S., Eastman, R., Kahn, R., Klein, R., Lachin, J., Lebovitz, H., Nathan, D., and Vinicor, F. (1998). Implications of the United Kingdom prospective diabetes study. *Diab. Care,* **21**: 2180–2184.

84. John, W.G. (1997). Glycated haemoglobin analysis. *Ann. Clin. Biochem,* **34**: 17–31.

85. Jeppsson, J.O., Kobold, U., Barr, J., Finke, A., Hoelzel, W., Hoshino, T., Miedema, K., Mosca, A., Mauri, P., Paroni, R., Thienpont, L., Umemoto,

M., and Weykamp, C. (2002). Approved IFCC reference method for the measurement of HbA. *Clin. Chem. Lab. Med*, **40**: 78–89.

86. Brunnekreeft, J.W.I., and Eidhof, H.H.M. (1993). Improved rapid procedure for simultaneous determinations of hemoglobins A1a, A1b, A1c, F, C, and S, with indication for acetylation or carbamylation by cation-exchange liquid chromatography. *Clin. Chem.*, **50**: 166–174.

87. Ellis, G., Diamandis, E.P., Giesbrecht, E.E., Daneman, D., and Allen, L.C. (1984). An automat of high pressure liquid chromatographic assay for hemoglobin A1c. *Clin. Chem.*, **30**: 1746–1742.

88. Turpeinen, U., Karjalainen, U., and Stenman, U.H. (1995). Three assays for glycohemoglobin compared. *Clin. Chem.*, **41**: 191–195.

89. Goldstein, D.E., Little, R.R., Lorenz, R.A., Malone, J.I., Nathan, D., Peterson, C.M., and Sacks, D.B. (2004). Tests of glycemia in diabetes. *Diabetes Care,* **27**: 1761–1773.

90. Lakshmy, R., and Gupta, R. (2009). Measurement of glycated hemoglobin A1c from dried blood by turbidimetric immunoassay. *J Diabetes Sci Technol*, **3**: 1203–1206.

91. Fluckiger, R., and Winterhalter, K.H. (1976). In vitro synthesis of hemoglobin. *FEBS Lett.*, **71**: 356–360.

92. Stöllner, D., Stöcklein, W., Scheller, F., and Warsink, E.A. (2002). Membrane-immobilized haptoglobin as affinity matrix for a hemoglobin-A1c immunosensor. *Anal. Chim. Acta*, **470**: 111–119.

93. Tsugawa, W., Ishimura, F., Ogawa, K., and Sode, K. (2000). Development of an enzyme sensor utilizing a novel fructosyl amine oxidase from a marine yeast. *Denki Kagaku Oyobi Kogyo Butsuri Kagaku,* **68**: 869–871.

94. Ogawa, K., Stollner, D., Scheller, F., Warsinke, A., Ishimura, F., Tsugawa, W., Ferri, S., and Sode, K. (2002). Development of a flow-injection analysis (FIA) enzyme sensor for fructosyl amine monitoring. *Anal. Bioanal. Chem.*, **373**: 211–214.

95. Sode, K., Ishimura, F., and Tsugawa, W. (2001). Screening and characterization of fructosyl-valine-utilizing marine microorganisms. *Mar. Biotechnol.*, **3**: 126–132.

96. Chawla, S., and Pundir, C.S. (2011). An electrochemical biosensor for fructosyl valine for glycosylated hemoglobin detection based on core-shell magnetic bionanoparticles modified gold electrode. *Biosens. Bioelectron.*, **26**: 3438–3443.

97. Fang, L., Li, W., Zhou, Y., and Liu, C.C. (2009). A single-use, disposable iridium-modified electrochemical biosensor for fructosyl valine for

the glycosylated hemoglobin detection. *Sens. Actuat. B Chem.,* **137**: 235–238.

98. Liu, S., Wollenberger, U., Katterle, M., and Scheller, F.W. (2006). Ferroceneboronic acid-based amperometric biosensor for glycated hemoglobin. *Sens. Actuat. B Chem.,* **113**: 623–629.

99. Chawla, S., and Pundir, C.S. (2012). An amperometric hemoglobin A1c biosensor based on immobilization of fructosyl amino acid oxidase onto zinc oxide nanoparticles-polypyrrole film. *Anal. Bioch.,* **430**: 156–162.

100. Krishna, V.S.R., Bhat, N., Bharadwaj, A., Chakrapani, K., and Srinivasan, S. (2011). Detection of glycated hemoglobin using 3-aminophenylboronic acid modified graphene oxide. IEEE/NIH *Life Science Systems and Applications Workshop* (LiSSA), 7/4/2011, Bangalore.

101. Xue, Q., Bian, C., Tong, J., Sun, J., Zhang, H., and Xia, S. (2011). A micro potentiometric immunosensor for hemoglobin-A1c level detection based on mixed SAMs wrapped nano-spheres array. *Biosens. Bioelectron.,* **26**: 2689–2693.

102. Xue, Q., Bian, C., Tong, J., Sun, J., Zhang, H., and Xia, S. (2011). CMOS and MEMS based micro hemoglobin-A1c biosensors fabricated by various antibody immobilization methods. *Sens. Actuat. A Phys.,* **169**: 282–287.

103. Xue, Q., Bian, C., Zhang, H., and Xia, S. (2010). An integrated micro immunosensor for hemoglobin-A1c level detection. In: *Mechatronics and Embedded Systems and Applications (MESA), 2010 IEEE/ASME International Conference on,* IEEE, Qingdao, pp. 208–212.

104. Kessler, G., and Lederer, H. (1966). Fluorimetric measurement of triglycerides. In: Kress, L.T. Jr. (ed.), *Automation in Analytical Chemistry,* Mediad Inc., New York, pp. 341–344.

105. Shukla, V.K.S. (1988). Recent advances in the high performance liquid chromatography of lipids. *Prog. Lipids Res.,* **27**: 5–38.

106. Ruiz-Gutiérrez, V., and Barron, L.J. (1995). Methods for the analysis of triacylglycerols. *J. Chromatogr. B Biomed. Appl.,* **671**: 133–168.

107. Byrdwell, W.C., Emkon, E.A., Neff, W.E., and Adlof, R.O. (1996). Quantitative analysis of triglycerides using atmospheric pressure chemical ionisation–mass spectrometry. *Lipids,* **31(9)**: 919–935.

108. Shireman, R.B., and Durieux, J. (1993). Microplate method for determination of serum cholesterol, high-density lipoproteins cholesterol, triglyceride and apolitroproteins. *Lipids,* **28(2)**: 151–155.

109. Werner, M., Gabrielson, D.G., and John, E. (1981). Ultramicro determination of serum triglycerides by bioluminescent assay. *Clin. Chem.,* **22(2)**: 268–271.

110. Björkhem, I., Blomstrand, R., and Svensson, L. (1976). Determination of serum triglycerides by mass fragmentograpy. *Clin. Chim. Acta*, **71**: 191–198.

111. Fossati, P., and Prencipe, L. (1982). Serum triglycerides determined colorimetrically with an enzyme that produces hydrogen peroxide. *Clin. Chem.*, **28**: 2077–2080.

112. Mendez, A.J., Cabeza, C., and Hsia, S.L. (1986). A fluorometric method for the determination of triglycerides in nanomolar quantities. *Anal. Biochem.*, **156**: 386–389.

113. Bacharik, P.S. and Woods, P.D.S. (1997). Laboratory considerations in the diagnosis and management of hyperlipoproteinemia, In: *Hyperlipidemia: Diagnosis and Therapy*, Rifkind, B.M. and Levy, R.I. (Eds), Grune and Stratton Foundation, New York, pp. 239–241.

114. McMaster, M.C. (2006). *HPLC: A Practical User's Guide*, Second Edition, John Wiley & Sons, Inc.

115. Narang, J., Chauhan, N., Rani, P., and Pundir, C.S. (2013). Construction of an amperometric TG biosensor based on AuPPy nanocomposite and poly(indole-5-carboxylic acid) modified Au electrode. *Bioproces. Biosys. Eng.*, **36**: 425–432.

116. Narang, J., Minakshi, Bhambi, M., and Pundir, C.S. (2010). Determination of serum triglyceride by enzyme electrode using covalently immobilized enzyme on egg shell membrane. *Int. J. Biol. Macromol.*, **47**: 691–695.

117. Narang, J., Bhambi, M., Minakshi, Pundir, C.S. (2010). Fabrication of an amperometric TG biosensor based on PVC membrane. *Anal. Lett.*, **43**: 1–10.

118. Narang, J., and Pundir, C.S. (2011). Construction of a triglyceride amperometric biosensor based on chitosan-ZnO nanocomposite film. *Int. J. Biol. Macromol*, **49**: 707–715.

119. Pundir, C.S., Sandeep Singh, B., and Narang, J. (2010). Construction of an amperometric triglyceride biosensor using PVA membrane bound enzymes. *Clin. Biochem.*, **43**: 467–472.

120. Pauliukaitea, R., Doherty, A. P., Murnaghan, K.D., and Bretta, C.M.A. (2011). Application of room temperature ionic liquids to the development of electrochemical lipase biosensing systems for water-insoluble analytes. *J. Electroanal. Chem.*, **656**: 96–101.

121. Narang, J., Chauhan, N., and Pundir, C.S. (2013). Construction of triglyceride biosensor based on nickel oxide-chitosan/zinc oxide/zinc hexacyanoferrate film. *Int. J. Biol. Macromol.*, **60**: 45–51.

122. Narang, J., Chauhan, N., Malhotra, N., and Pundir, C.S. (2014). Fabrication of triglyceride biosensor based on magnetic nanoparticles/zinc oxide/zinc hexacyanoferrate film: Novel immobization matrix for electrochemical sensing. *Adv. Sci. Lett.*, **20**: 1331–1336.

123. Winartasaputra, H., Kutan, S.S., and Cuilbault, G.C. (1982). Amperometeric enzyme determination of triglyceride in serum. *Anal. Chim*, **54**: 1987–1990.

124. Compagnone, D., Esti, M., Messia, M.C., Peluso, E., and Palleschi, G. (1998). Development of a biosensor for monitoring of glycerol during alcoholic fermentation. *Biosens. Bioelectron*, **13**: 875–880.

125. Minakshi, and Pundir, C.S. (2008). Construction of an amperometric enzymic sensor for triglyceride determination. *Sens. Actuat. B Chem.*, **133**: 251–255.

126. Dhand, C., Solanki, P.R., Datta, M., and Malhotra, B.D. (2010). Polyaniline/Single-walled carbon nanotubes composite based triglyceride biosensor. *Electroanalysis*, **22**: 2683–2693.

127. Velmurugan, M., Sakthinathan, S., Chen, S.M., and Karuppiah, C. (2015). Direct electron transfer of glucose oxidase and electrocatalysis of glucose based on gold nanoparticles/electroactivated graphite nanocomposite. *Int. J. Electrochem. Sci.*, **10**: 6663–6671.

128. Aulenta, F., Rossetti, S., Amalfitano, S., Majone, M., and Tandoi, V. (2013). Conductive magnetite nanoparticles accelerate the microbial reductive dechlorination of trichloroethene by promoting interspecies electron transfer processes. *ChemSusChem*, **6**: 433–436.

129. Lee, J., Shim, H.S., Lee, M., Song, J.K., and Lee, D. (2011). Size-controlled electron transfer and photocatalytic activity of ZnO–Au nanoparticle composites. *J. Phys. Chem. Lett.*, **2**: 2840–2845.

130. Anta, J.A. (2012). Electron transport in nanostructured metal-oxide semiconductors, *Cur. Op. Col. Interf.*, **17**: 124–131.

131. Stephen, R.B., Fallyn, W.C., Edmund, J.F.D., and Richard, G.C. (2010). Nanoparticle-modified electrodes *Phys. Chem. Chem. Phys.*, **12**: 11208–11221.

132. Solanki, P.R., Dhand, C., Kaushik, A., Ansari, A., Sood, A K.N., and Malhotra, B.D. (2009). Nanostructured cerium oxide film for triglyceride sensor. *Sens. Actuat. B Chem.*, **141**: 551–556.

133. Liao, W.Y., Liu, C.C., and Chou, T.C. (2008). Detection of triglyceride using an iridium nano-particle catalyst based amperometric biosensor. *Analyst*, **133**: 1757–1763.

134. Ganjali, M.R., Faridbod, F., Nasli-Esfahani, E., Larijani, B., Rashedi, and Norouzi, H.P. (2010). FFT continuous cyclic voltammetry triglyceride

dual enzyme biosensor based on MWCNTs-CeO$_2$ nanoparticles. *Int. J. Electrochem. Sci.*, **5**: 1422–1433.

135. Ansari, A.A., Kaushik, A., Solanki, P.R., and Malhotra, B.D. (2008). Sol–gel derived nanoporous cerium oxide film for application to cholesterol biosensor. *Electrochem. Commun.*, 10: 1246–1249.

136. Ansari, A.A., Solanki, P.R., and Malhotra, B.D. (2008). Sol-gel derived nanostructured cerium oxide film for glucose sensor. *Appl. Phys. Lett.*, **92**: 263901–263903.

137. Nakako, M., Hanazato, Y., Maeda, M., and Shiono, S. (1986). Neutral lipid enzyme electrode based on ion-sensitive field effect transistors, *Anal. Chim. Acta*, **185**: 179–185.

138. Setzu, S., Salis, S., Demontis, V., Salis, A., Monduzzi, M., and Mula, G. (2007). Porous silicon-based potentiometric biosensor for triglycerides, *Physica. Status. Solid*, **204**: 1434–1438.

139. Wilhelm, D., Voigt, H., Treichel, W., Ferretti, R., and Prasad, S. (1991). pH sensor based on differential measurements on one pH-FET chip, *Sens. Actuat. B.*, **4**: 145–149.

140. Reddy, R.R.K., Chadha, A., and Bhattacharya, E. (2001). Porous silicon based potentiometric triglyceride biosensor. *Biosens. Bioelectron.*, **16**: 313–317.

141. Basu, I., Subramanian, R.V., Mathewa, A., Kayasthac, A.M., Chadha, A., and Bhattacharya, E. (2005). Solid state potentiometric sensor for the estimation of tributyrin and urea, *Sens. Actuat. B Chem.*, **107**: 418–423.

142. Pijanewska, D.G., Baraniecka, A., Wiater, R., Ginalska, G., Lobarzewski, J., and Torbicz, W. (2001). The pH-detection of triglycerides. *Sens. Actuat. B Chem.*, **78**: 263–266.

143. Vijayalakshmi, A., Tarunashree, Y., Baruwati, B., Manorama, S.V., Narayana, B.L., Johnson, R.E.C., and Rao, N.M. (2008). Enzyme field effect transistor (ENFET) for estimation of triglycerides using magnetic nanoparticles. *Biosens. Bioelectron.*, **23**: 1708–1714.

144. Fernandez, R.E., Vemulachedu, H., Bhattacharya, E., and Chadha, A. (2009). Comparison of a potentiometric and a micromechanical triglyceride biosensor. *Biosens. Bioelectron.*, **24**: 1276–1280.

145. Wyss, M., and Kaddurah-Daouk, R. (2000). Creatine and creatinine metabolism. *Physiol. Rev.*, **80**: 1107–1213.

146. Miller, B.F., and Dubos, R. (1937). Determination by a specific enzymatic method of the creatinine content of blood and urine from normal and nephritic individuals. *J. Biol. Chem.*, **121**: 457–464.

147. Sadilek, L. (1965). Creatinine determination in the urine of alcaptonuric patients. *Clin. Chirn. Act.,* **12**: 436–439

148. Takatsu, A., and Nishi, S. (1993). Determination of serum creatinine by isotope dilution method using discharge-assisted thermospray liquid chromatography/mass spectrometry. *Biol. Mass Spectrom.,* **22**: 643–646.

149. Siekmann, L. (1985). Determination of creatinine in human serum by isotope dilution-mass spectrometry. *J. Clin. Chem. Clin. Biochem.,* **23**: 137–144.

150. Fridolin, I., Jerotskaja, J., Lauri, K., Uhlin, F., and Luman, M. (2010). A new optical method for measuring creatinine concentration during dialysis. *IFMBE Proc.,* **29**: 379–382.

151. Tomson, R., Uhlin, F., Holmar, J., Lauri, K., Luman, M., and Fridolin, I. (2011). Development of a method for optical monitoring of creatinine in the spent dialysate. *Eston. J. Eng.,* **17**: 140–150.

152. Suzuki, H., Arakawa, H., and Karube, I. (2001). Fabrication of a sensing module using micromachined biosensors. *Biosens. Bioelectron.,* **16**: 725–733.

153. Tsuchida, T., and Yoda, K. (1983). Multi-enzyme membrane electrodes for determination of creatinine and creatine in serum. *Clin. Chem,* **29**: 51–55.

154. Osborne, M.D., and Girault, H.H. (1995). The micro water/1,2-dichloroethane interface as a transducer for creatinine assay. *Mikrochim. Acta,* **117**: 175–185.

155. Choi, S.H., Lee, S.D., Shin, J.H., Ha, J., Nam, H., and Cha, G.S. (2002). Amperometric biosensors employing an insoluble oxidant as an interference-removing agent. *Anal. Chim. Acta,* **461**: 251–260.

156. Kubo, I., and Karube, I. (1986). Immobilization of creatinine deiminase on a substituted poly(methylglutamate) membrane and its use in a creatinine sensor. *Anal. Chim. Acta,* **187**: 31–37.

157. Nguyen, V.K., Wolff, C.M., Seris, J.L., and Schwing, J.P. (1991). Immobilized enzyme electrode for creatinine determination in serum. *Anal. Chem.,* **63**: 611–614.

158. Rui, C.S., Ogawa, H.I., Sonomoto, K., and Kato, Y. (1993). Multifunctional flow-injection biosensor for the simultaneous measurement of creatinine, glucose and urea. *Biosci. Biotech. Biochem.,* **57**: 191–194.

159. Rui, C.S., Sonomoto, K., and Kato, Y. (1992). Amperometric flow-injection biosensor system for the simultaneous determination of urea and creatinine. *Anal. Sci.,* **8**: 845–850.

160. Rui, C.S., Sonomoto, K., Ogawa, H.I., and Kato, Y. (1993). A flow-injection biosensor system for the amperometric determination of creatinine: Simultaneous compensation of endogenous interferents. *Anal. Biochem.*, **210**: 163–171.

161. Yamato, H., Ohwa, M., and Wemet, W. (1995). A polypyrrole/three-enzyme electrode for creatinine detection. *Anal. Chem.*, **67**: 2776–2780.

162. Madaras, M.B., Popescu, I.C., Ufer, S., and Buck, R.P. (1996). Microfabricated amperometric creatine and creatinine biosensors. *Anal. Chim. Acta*, **319**: 335–345.

163. Madaras, M.B., and Buck, R.P. (1996). Miniaturized biosensors employing electropolymerized permselective films and their use for creatinine assays in human serum. *Anal. Chem.*, **68**: 3832–3839.

164. Schneider, J., Grhdig, B., Renneberg, R., Cammann, K., Madaras, M.B., Buck, R.P., and Vorlop, K.D. (1996) Hydrogel matrix for three enzyme entrapment in creatine/creatinine amperometric biosensing. *Anal. Chim. Acta*, **325**: 161–167.

165. Shih, Y.T., and Huang, H.J. (1999). A creatinine deiminase modified polyaniline electrode for creatinine analysis. *Anal. Chim. Acta*, **392**: 143–150.

166. Erlenkotter, A., Fobker, M., and Chemnitius, G.C. (2002). Biosensors and flow-through system for the determination of creatinine in hemodialysate. *Anal. Bioanal. Chem.*, **372**: 284–292.

167. Yadav, S., Devi, R., Bhar, P., Singhla, S., and Pundir, C.S. (2012). A creatinine biosensor based on iron oxide nanoparticles/chitosan-g-polyaniline composite film electrodeposited on Pt electrode. *Enz. Microb. Technol*, **50**: 247–254.

168. Yadav, S., Devi, R., Kumar, A., and Pundir, C.S. (2011). Tri-enzyme functionalized ZnO-NPs/CHIT/c-MWCNT/PANI composite film for amperometric determination of creatinine. *Biosens. Bioelectron.*, **28**: 64–70.

169. Gutierrez, M., Alegret, S., and del Valle, M. (2008). Bioelectronic tongue for the simultaneous determination of urea, creatinine and alkaline ions in clinical samples. *Biosens. Bioelectron.*, **23**: 795–802

170. Tiwari, A., and Dhakate, S.R. (2009). Chitosan–SiO_2–multiwall carbon nanotubes nanocomposite: A novel matrix for the immobilization of creatine amidinohydrolase. *Int. J. Biol. Macromol.*, **44**: 408–412.

171. Tiwari, A., and Shukla, S.K. (2009). Chitosan-g-polyaniline: A creatine amidinohydrolase immobilization matrix for creatine biosensor. *eXPRESS Poly. Lett.*, **3**: 553–559.

172. Pookaiyaudom, P., Seelanan, P., Lidgey, F.J., Hayatleh, K., and Toumazou, C. (2011). Measurement of urea, creatinine and urea to creatinine ratio using enzyme based chemical current conveyor (CCCII+). *Sens. Actuat. B Chem.*, **153**: 453–459.

173. Yadav, S., Devi, R., Kumari, S., Yadav, S., and Pundir, C.S. (2011). An amperometric oxalate biosensor based on sorghum oxalate oxidase bound carboxylated multiwalled carbon nanotubes–polyaniline composite film. *J. Biotechnol.*, **151**: 212–217.

174. Chen, C.H., and Lin, M.S. (2012). A novel structural specific creatinine sensing scheme for determination of urine creatinine. *Biosens. Bioele.*, **31**: 90–94.

175. Kumar, N.T., Ananthi, A., Maithyarasu, J., Joseph, J., Phani, K.L., and Yegnaraman, V. (2011). Enzymeless creatinine estimation using poly(3,4-ethlenedioxythiophene)-β-cyclodextrin. *J. Electroanl. Chem.*, **661**: 303–308.

Chapter 7

Microfluidics: A Platform for Futuristic Sensors

Ashish Mathur and Shikha Wadhwa

Amity Institute of Nanotechnology, Amity University, Noida-201313, India
amathur@amity.edu

7.1   Introduction

Microfluidics is the science dealing with the design, fabrication, and formulation of devices and processes to deal with very small volumes (i.e., nanoliters) of fluid. The device dimensions range from millimeters down to micrometers. Therefore, the fabrication processes for such devices differ greatly from that on the macroscale. When the dimensions of the device are comparable to the size of the particle of the fluid that passes through it, the system behavior is affected. In general, the laws governing the fluid flow in the microenvironment are the same as those describing the flow at the macroscale. Miniaturization of the device dimensions adds some characteristic features that can be influenced to carry out processes that are not possible at the macroscale. In microfluidic devices, most of the physical characteristics, for example, surface-area-to-volume ratio, surface tension, and diffusion, do not just scale

Biosensors: An Introductory Textbook
Edited by Jagriti Narang and Chandra Shekhar Pundir
Copyright © 2017 Pan Stanford Publishing Pte. Ltd.
ISBN 978-981-4745-94-9 (Hardcover), 978-1-315-15652-1 (eBook)
www.panstanford.com

linearly from macro- to micro-devices. Some common fluids that are used in microfluidics include protein or antibody solutions, buffer solutions, bacterial or cell suspensions, and blood samples. There are numerous significant advantages of using microfluidic devices to conduct biomedical research over other methods:

- Smaller amounts of reagents are required to obtain a diagnosis since small device dimensions are involved.
- Cost-effective and easily scaled-up processes are available for the fabrication of microfluidic devices.
- Highly complex lab-on-a-chip device fabrication can be achieved for rapid diagnosis, which eliminates the requirement for laboratory analysis.

The fluid flow in microfluidics is based on the classical theories of fluid dynamics: low–Reynolds number flows. Fluid flow at the microlevel in a microfluidic channel is defined as having a low Reynolds number (<100).

$$R_e = \frac{\rho v_s^2 / L}{\mu v_s / L^2} = \frac{\rho v_s L}{\mu} = \frac{v_s L}{v} = \frac{\text{Inertial forces}}{\text{Viscous forces}} \tag{7.1}$$

The Reynolds number, as given in Eq. 7.1, defines the fluid flow and is dependent on various factors such as viscosity, fluid density, relative length scale, and average velocity. A low Reynolds number indicates minimal turbulence, implying that the flow occurs in a fairly predictable manner. The movement of a fluid through a microfluidic system is facilitated by capillary forces. A fluid flows when the adhesive molecular forces are stronger than the cohesive intermolecular forces present in the fluid. Nevertheless, this model may be an oversimplified one. Due to the appearance of smaller devices, many different variables have been introduced, such as:

- Surface characteristics, including surface tension, which in turn is affected by surface roughness, electrical effects, and van der Waals forces.
- Much complex three-dimensional patterning of the surface of the microfluidic device.
- The occurrence of suspended particles having size similar to that of the dimensions of the device.

An important challenge in the development of such a device is understanding the fluid behavior at the microlevel, which is governed by complex variables, and utilizing and manipulating them to one's advantage. On the developed device, the fluid flow through the system is governed by two different forces. The force of gravity will allow the drop to enter the capillary channel at one end, while the capillary action will draw the fluid through the system.

7.2 Microfluidic Basics

7.2.1 General Microfluidic Theory

Lammertyn et al. [1] have developed a convection–diffusion–reaction model to mimic the behavior of a flow inject analysis biosensor as a function of flow rate, fluidic channel dimensions, flow cell geometry, volume of substrate, and the fluid used. This work investigates the fluid flow behavior using Navier–Stokes equations and Michaelis–Menten kinetics. On the basis of the velocity flow profile obtained, pressure-driven fluidics and standard flow were compared. The modeling approach followed in this work has a high potential in the design and development of biosensors with high accuracy, irrespective of the enzyme chosen.

Kuswandi et al. [2] reviewed the application of optical sensing systems in microfluidic devices. Both off-chip (macroscale optical infrastructure coupled externally to the device) and on-chip approaches (which comprises the integration of micro-optical functions into the microfluidic device) have been considered in detail. The shape and size of the sensor demonstrate a profound effect on the detection limits due to analyte transport limitation. The review concludes with an assessment of the future directions of on-chip integrated optical sensing microfluidic devices.

In another work, Gould et al. [3] discussed the commercial potential of microfluidic devices. Microfluidic lab-on-a-chip devices can be standalone diagnostic devices as they can contain everything required to probe the smallest of liquid samples in their micro-dimensional channels for diagnosis. The devices can be active (requiring external control) or passive (completely self-contained). These items are one-time devices, which are designed to be discarded after single use. Therefore, they must be cheaper and easier to fabricate. This work underlines the etching and lithography techniques for microfluidic device fabrication. It also highlights the devices that can be integrated with the design of electronic devices to manipulate an electric field and affect fluid movement as well as process nanoliter samples through capillary circuitry. In addition, the work also mentions the area of nanofluidics, which uses dimensions comparable to the size of the molecules in question. The manipulation of these molecules and squeezing them through the channel using key parameters such as light at certain wavelengths will enable one to distinguish between different molecular types.

Woias et al. [4] reviewed the advancement of microfluidics and foresighted its future prospects. The review deals with the micropump area of microfluidics specifically and analyzes the potential futuristic applications of microfluidic micropumps such as biochemical sensing. Woias visions the continuation of the advancement of microfluidics, which will fuel the development of micropump systems in the near future.

Stone et al. [5] overviewed the key areas of microfluidics and discussed the physics and various variables involved in the fabrication of such a device. They emphasized that many microfluidic devices have been already developed over the past few years and such systems are becoming more and more imperative to the biomedical and pharmaceutical industries, and other areas outside these two main research realms. The skill to implement patterning on such a small scale, fabricate a device based on the lab-on-a-chip concept, manipulate, control, and augment chemical reactions, as well as develop mixing and separation processes will, no doubt, pave new ways for research opportunities, eventually leading to commercial opportunities for the successful devices. The areas that require further research have also been highlighted in the review. These areas include molecular interactions, surface forces, and fluid flow.

Gervais et al. [6] looked upon mass transport and surface reactions in microfluidic systems. Their work deals with the analysis of diffusion and laminar flow convection when combined with surface reactions, which are relevant to microchemical assays. It also compares analytic solutions for the concentration fields with the predictions from two-dimensional computer models that are commonly used to interpret such results. Particular emphasis is on the characterization of transport in shallow microfluidic channels. The key parameters, capture fraction of the bulk analyte and the saturation timescale at the reactive surface, which are significant to onboard chip biochemical assays and microfluidic sensors, were investigated and compiled.

7.2.2 Production Techniques

Pepin et al. [7] developed microfluidic devices for biomolecule separation using an array of well-defined nanostructures. Based on the processing methods, two types of pattern replication of the same device configuration have been considered. In the first approach, a tri-layer nanoimprint lithography is carried out to pattern a SiO_2 substrate comprising a plastic cover plate on top. The second method involves directly imprinting thermoplastic polymer pellets to form bulk plates, which were then thermally bonded together. The devices are characterized by epifluorescence microscopy to track leaks and fluid penetration. The possibility of fabricating nanofluidic devices and their mass production are the key findings of the work.

Wu and Chen [8] developed a novel method for fabricating a microfluidic structure on a polymer substrate. The method uses an optical disc process to avoid damaging the mirror plate of the mold. By means of a new cooling system, the cycle time has been significantly reduced in comparison to the conventional methods. The molding system consists of a stamper and a vacuum system, which joins the mold insert with the mold. Therefore, the time to change the mold is drastically reduced. It also reduces the time needed from hours to only a few minutes. The method has been found to be suitable for mass production.

Ruano-López et al. [9] developed a microfluidic device using SU-8 as a substance for the first time. SU-8 is a photo-definable epoxy, which has been used in this work to integrate optical sensors and microfluidic structures. The chip consists of optical sensors, waveguides, and sealed microfluidic channels patterned in SU-8 on a silicon substrate. Low-temperature adhesive bonding of the SU-8 patterned films at a wafer level enables sealing of microchannels. This fabrication process has many advantages; it is fast, reproducible,

CMOS compatible, and an easy method to develop a lab-on-a-chip device.

Sun et al. [10] established a new method for the fast prototyping of hard polymer microfluidic systems using solvent imprinting and bonding. For solvent imprinting, a layer of SU-8 photoresist is patterned on glass as a template. The first step is the exposure of poly(methyl methacrylate)(PMMA) to acetonitrile followed by the pressing of SU-8 template into the surface, which provides suitably imprinted channels and a proper surface for bonding. In a similar manner, a PMMA cover plate is fabricated and bonded together at room temperature and at an appropriate pressure. The fabrication is accomplished in 15 min, and the SU-8 template can be used to produce around 30 PMMA chips. The repeatability of the fabrication process is also established.

Yu et al. [11] proposed an adhesive bonding technique at wafer level again using SU-8 as the structural material. The adhesive is imprinted onto one of the surfaces with the aim to bond the two surfaces using a low-temperature method. The overall process involves three steps; first, the adhesive layer is deposited onto the bonding surface by contact imprinting onto the SU-8 photoresist. In the second step, the wafers are placed in contact and aligned. The final step involves the bonding of the two surfaces performed at 100–200°C at a pressure of 1000 N in a vacuum. This process is verified in the fabrication process of a dielectrophoretic device.

Bilenberg et al. [12] investigated an adhesive bonding technique for the wafer level sealing of SU-8 based lab-on-a-chip microsystems. Microfluidic channels are created using a standard lithography process in SU-8 photoresist and sealed with a Pyrex glass lid by means of an intermediate layer of PMMA. This bonding technique has been compared with SU-8 bonding, and it has been found that the slow flow of SU-8 resist during the sealing process caused the channels to be filled with resist. The bonding strengths of both methods have been tested, and it has been found that PMMA has a bonding strength of around 16 MPa, when bonded under a force of 2000 N and a temperature of 120°C.

Liu et al. [13] demonstrated the fabrication of microchannels on PMMA substrates using novel microfabrication techniques. The image of microchannels is transferred from a silicon master plate using hot embossing methods. The silicon master is electrostatically bonded to a Pyrex glass wafer, which enhances the device yield from 20 per master to over 100.

Poenar et al. [14] reported on the fabrication of a microfluidic device from glass substrates for characterizing cells using impedance spectroscopy. The device is constructed from two glass wafers. The bottom wafer contained microfluidic channels and electrodes, while the upper wafer had inlets and outlets. The major focus of this paper is the fabrication of the device involving, first, the application of through-wafer wet etching to pattern inlets and outlets on the lid wafer and, second, patterning of the electrodes in the microfluidic channel on the bottom wafer. Finally, bonding is carried out using an intermediate bonding layer (paralyene C), which requires a low bonding temperature, short time period for bond to take effect, and high bond strength.

7.3   Importance of Microfluidics

Among the various advantages of microfluidics, one apparent advantage is that the reduced size of components and processes uses lesser volumes of fluid, therefore leading to reduced reagent consumption. Hence, the costs drop down significantly and allow small quantities of samples to be stretched further (for example, multiple screening assays). This way, the volume of waste products is also decreased. Rapid heat transfer is facilitated by the low thermal mass and large surface-to-volume ratio of miniaturized components. This enables fast temperature changes and precise control over temperature. This feature leads to the elimination of the build-up of heat or "hot spots" in an exothermic reaction, thereby eliminating the formation of undesired side products or even explosions. In processes relating support-bound catalysts/enzymes and in solid-phase synthesis, a large surface-to-volume ratio is of great benefit. At such small dimensions of microfluidic devices, diffusive mixing is rapid, which increases the speed and precision of reactions. Significant improvements in performance reflected in enhanced sensitivity, higher selectivity, and better repeatability are frequently seen in microfluidic assays due to the reduced measurement times [15].

In view of the above discussion, the study of microfluidics becomes significant because of the following reasons:

- Reduction in the size of the device
- Handling of less amount of fluids
- Increase in the speed of reaction
- Reduced consumption of chemical reagents
- Reduction in the cost of reagents
- Higher surface-to-volume ratio/low Reynolds number
- Safety

The advantages of microfluidic devices are summarized as follows:

- Can work with small volume
- Lower power consumption and better performance
- Can be integrated with other devices—lab on a chip
- Ease of disposing of devices and fluids
- Portability
- Reliability
- User friendly
- Minimized size of chip

7.4 Manufacturing Methods

In the early days, the fabrication of microfluidic devices mainly relied on techniques transferred from the conventional two-dimensional integrated circuit (IC) and silicon-based two- or three-dimensional MEMS processes. This includes photolithography, thin film metallization, and chemical etching. Later, glass-based, glass-silicon, glass-polymer mixed microfluidic fabrication techniques and devices started to garner more interest [16–26]. The glass materials were preferred partly for the biocompatibility toward biomedical applications [27–31] and for the ideal surface characteristics where high temperature or strong solvents should appear [32–40] (e.g., on-chip capillary electrophoresis-based operations). However, the lack of optical transparency at interested wavelengths (for silicon), micromachining difficulties, and comparably high expenses for both silicon and glass material shave hampered their wider applications in microfluidics. Tremendous efforts have been made to find alternative materials that are more cost-effective and easier for micromachining. With the development of related fabrication techniques in the recent years, the polymer/plastic-based microfluidic systems have garnered more interest than their conventional competitors [41]. In spite of comparatively weak bonding and structure deformation during device packaging processes, polymer materials still seem attractive due to the facts that they are more economic compared with silicon and glasses, are easier to be fabricated in/on, avoid high-temperature annealing and stringent cleaning, and are more system integration friendly (e.g., interconnections). Moreover, there exists a wider range of materials to be chosen for characteristics that are required for each specific application, such as good optical transparency, biocompatibility, and chemical or mechanical properties. Another important reason for the interest from both academia and industry on polymer microfluidic devices is the possibility of disposable microfluidic chips toward biomedical and clinical applications. These devices usually require low cost of fabrication, high volume production, good reproducibility, and versatility in design for a wide spectrum of specific applications. The current methods for the fabrication of microfluidic devices include prototyping techniques [42–44] and direct fabrication techniques such as laser photoablation or laser micromachining, photolithography/optical lithography, and X-ray lithography. To date, most of the current soft lithography processes still rely on modern photolithography techniques for master template/mask fabrication. Consequently, the low resolution ability of soft lithography can be gradually improved with the high-quality masks by modern photolithography. Sub 100 nm fabrication resolution can also be achieved by composite layers of stamps. Other techniques have also been used to obtain soft lithography masters with nanometer-scale features below 5 nm, such as to replicate those features from single-walled carbon nanotubes or from crystal fractures as soft lithography masters. For photolithography made masters for the soft lithography process, the recently reported resolution limit has been pushed to around 20 nm.

Fabrication of a device largely depends on the material selection. Most common materials that are used for microfluidic chip manufacturing are low fluorescence Schott Borofloat glass, Corning borosilicate glass, fused silica, quartz silicon, PMMA, SU-8, PDMS, steel, aluminum, and copper [45–48].

7.4.1 Polymer-Based Micromachining Techniques

Polymeric micromachining uses polymers as structural material. The most well-known polymeric technique is LIGA (a German acronym for *Lithographie, Galvanoformung, Abformung* or Lithography, Electroplating, and Molding), which includes thick resist photolithography, electroplating, and micro-molding. Besides the conventional X-ray lithography, there are a number of alternative polymeric techniques for high-aspect-ratio structures. This section discusses each of these techniques separately:

- Thick resist lithography
- Polymeric bulk micromachining
- Polymeric surface micromachining
- Micro-stereo lithography
- Micro-molding

7.4.1.1 Thick resist lithography

7.4.1.1.1 *Poly(methyl methacrylate) resist*

PMMA was originally used as a resist material for the LIGA technique [49]. The material is well known by a variety of trade names such as Acrylic, Lucite, Oroglas, Perspex, and Plexiglas. PMMA can be applied on a substrate by different methods: multiple spin coating, prefabricated sheets, casting, and plasma polymerization. Multilayer spin coating causes high interfacial stresses, which lead to cracks in the resist layer. These cracks can be avoided by using a preformed PMMA sheet, which is bonded to the substrate [50]. Monomer MMA (methyl methacrylate) can be used as the adhesive material for the bonding process [51]. PMMA can also be polymerized in situ with casting resin [52] or with plasma [53]. Structuring PMMA requires collimated X-rays with wavelengths ranging from 0.2 to 2 nm, which are only available in synchrotron facilities. X-rays also require special mask substrates such as beryllium and titanium, which further increases the cost of this technique. The beryllium mask with its higher Young's modulus and thickness is optimal for X-ray lithography. The absorber material of an X-ray mask can be gold, tungsten, or tantalum. The thicker the absorber layer, the higher the X-ray energy and the higher the aspect ratio in PMMA. X-rays change the PMMA property in the exposed area, which is chemically etched in the development process. The developer consists of a mixture of 20 vol% tetrahydro-1,4-oxazine, 5 vol% 2-aminoethanol-1, 60 vol% 2-(2-butoxy-ethoxy) ethanol, and 15 vol% water [54]. The limited access and high costs of a synchrotron facility are the main drawbacks of the LIGA technique, in general, and PMMA as a polymeric structural material, in particular. Thick film resist such as SU-8 and the AZ-4000 series has the advantage of using low-cost UV exposure. However, structure heights and aspect ratios of UV exposure cannot meet those of PMMA with X-ray exposure.

7.4.1.1.2 *SU-8 resist*

SU-8 is a thick epoxy-based negative photoresist (Fig. 7.1). It is a very viscous polymer that can be spun or spread over a thickness ranging from 0.1 μm to 2 mm and still be processed with standard lithographic techniques. It can be used to pattern high–aspect ratio (>20) structures. Its maximum absorption is for ultraviolet light with a wavelength of 365 nm. When exposed, SU-8's long molecular chains cross link causing the solidification of the material. Once hard bake, SU-8 poses excellent mechanical properties and can be used for making moveable parts.

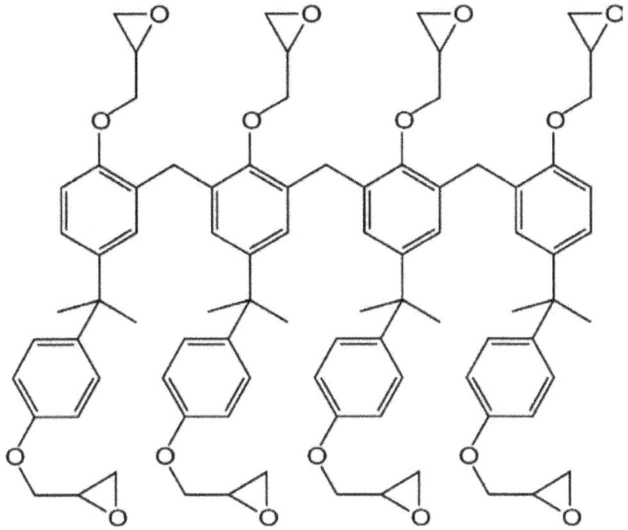

Figure 7.1 A chemical structure of SU-8 molecule.

It is also one of the most biocompatible materials known and is often used in bio-MEMS. SU-8 is highly transparent in the ultraviolet region, allowing fabrication of relatively thick (hundreds of micrometers) structures with nearly vertical side walls. After exposition and developing, its highly cross-linked structure gives it high stability to chemicals and radiation damage. Cured cross-linked SU-8 shows very low levels of outgassing in a vacuum. However, it is very difficult to remove and tends to outgas in an unexposed state.

SU-8 is a negative photoresist based on EPON SU-8 epoxy resin for the near-UV wavelengths from 365 to 436 nm. At these wavelengths, the photoresist has very low optical absorption, which makes photolithography of thick films with high aspect ratios possible. This resist was developed by IBM [55, 56]. The material was adapted for MEMS applications during collaboration between the EPFL Institute of Microsystems and IBM-Zurich [57, 58]. Structure heights up to 2 mm with an aspect ratio better than 20 can be achieved with standard lithography equipment. Photoresists such as SU-8 are based on epoxies, which are referred to as oxygen bridges between two atoms. Epoxy resins are molecules with one or more epoxy groups. During the curing process, epoxy resins are converted to a thermoset form or a three-dimensional network structure. SU-8 photoresist consists of three basic components:

1. Epoxy resin such as EPON SU-8
2. Solvent, called gamma-butyrolactone (GBL)
3. Photoinitiator, such as triarylium-sulfonium salts

7.4.1.1.3 *Other thick film resists*

AZ9260 is the other Novolak photoresist from Clariant. AZ9260 exhibits a better transparency compared to AZ4562 and, therefore, promises a better aspect ratio. Aspect ratios up to 15 are achieved with a film thickness of 100 μm [59]. A theoretical thickness of 150 μm is expected from this photoresist. Ma-P 100 (Microresist Technology, Berlin, Germany) is the other photoresist that can give structure heights up to 100 μm. This photoresist has aspect ratios on the order of 5, poorer than that of the AZ-family [60].

7.4.1.1.4 *Polyimide*

Polyimide is commercially available as photoresists such as Proimide 348 or 349 (Ciba Geigy) or PI-2732 (DuPont). A single spin can result in a film thickness of up to 40 μm. Photosensitive polyimide can be used for the same purpose as other thick resists described in the previous sections [61]. Fluorinated polyimide is an interesting material because of its optical transparency and simple machining. In RIE processes of this material, fluorine radicals are released from the fluorinated polyimide and act as etchants [62]. Polyimide is a good substrate material. Metals such as aluminum, titanium, and platinum can be sputtered on it using the lift-off technique [63]. Similar to other polymers, polyimide can be etched with RIE in oxygen plasma. Combining photolithography, RIE, and lamination, complex channel structures with metal electrodes can be fabricated in polyimide [64].

7.4.1.1.5 *Parylene*

Parylene is a polymer that can be deposited with chemical vapor deposition (CVD) at room temperature. The CVD process allows coating a conformal film with a thickness ranging from several microns to several millimeters. The basic type is parylene N, which is poly-paraxylylene. Parylene N is a good dielectric, exhibiting a very low dissipation factor, high dielectric strength, and a frequency-independent dielectric constant.

Parylene C is produced from the same monomer, modified only by the substitution of a chlorine atom for one of the aromatic hydrogens. Parylene C has a useful combination of electrical and physical properties as well as a very low permeability to moisture and other corrosive gases. Parylene C can also provide a conformal insulation. Parylene D is modified from the same monomer by the substitution of the chlorine atom for two of the aromatic hydrogens. Parylene D is similar in properties to parylene C with the added ability to withstand higher temperatures. Figure 7.2 shows the chemical structures of three types of parylene discussed in this section. Deposition rates are fast, especially for parylene C, which is normally deposited at a rate of about 10 μm/min. The deposition rates of parylene N and parylene D are slower. Parylene can be used in microfluidic devices as a structural material, which offers low Young's modulus. Such a soft material is needed in microvalves and micropumps. Furthermore, parylene coating can improve the biocompatibility of a microfluidic device.

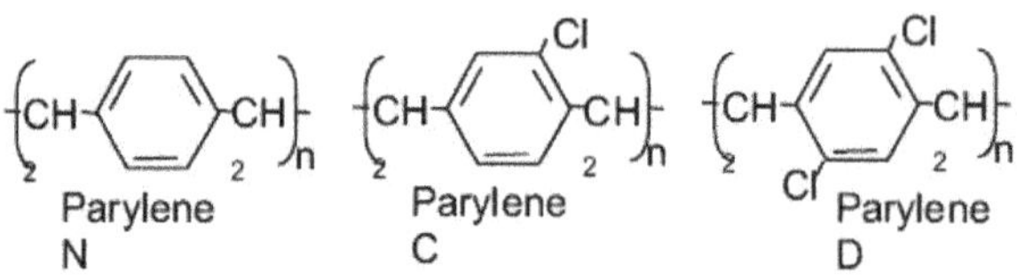

Figure 7.2 Chemical structures of parylene.

7.4.1.1.6 *Electro-depositable photoresist*

Eagle ED2100 and PEPR 2400 (Shipley Europe Ltd., England) are electrodepositable photoresists. These photoresists were originally developed for printed circuit boards. The photoresist is an aqueous emulsion consisting of polymer micelles. The resist is deposited on wafers by a cataphoretic electrodeposition process [65]. In an electric field, positively charged micelles move to the wafer, which works as a cathode. The polymer micelles coat the wafer until the film is so thick that the deposition current approaches zero. The nominal resist thickness is about 3 to 10 µm. The thickness and aspect ratios of electrodepositable photoresists are not relevant for thick film application. Electrodepositable photoresist is suited for applications where it is difficult to spin on a resist after patterning the substrate surface. Due to the relatively conformal nature of the deposition process, electrodepositable photoresists are interesting sacrificial materials for polymeric surface micromachining.

7.4.1.1.7 *Conductive polymers*

Conductive polymers or conjugated polymers are polymeric materials, which have received the growing attention of the MEMS community. Conjugated polymers have alternating single and double bonds between carbon atoms along the polymer backbone. The conjugation results in a band gap and makes the polymers behave as semiconductors. Conjugated polymers can be doped electrochemically, electrically, chemically, or physically with ion implantation. In the doped state, conjugated polymers are electrically conducting. These characteristics allow conductive polymers to be used as the material for electronic devices such as diodes, light-emitting diodes, and transistors. Furthermore, the doping level of polymers is reversible and controllable. In some conducting polymers, the change of doping level leads to volume change, which can be used as an actuator. The most common and well-researched conjugated polymer is polypyrrole (PPy). The major deposition techniques for conjugated polymers are the following.

- In *spin coating of polymers*, the polymer is dispersed or dissolved in a solvent. The resin can be spin coated on a substrate. Melted polymers can also be applied directly on the substrate.
- In *spin coating of precursor and subsequent polymerization*, precursor polymers are dissolved in a solvent. The polymerization occurs at an elevated temperature. Polymerization can also be achieved by a UV-cross link of the precursor with a photoinitiator, similar to the developing process of SU-8.
- *CVD* is similar to the technique used for parylene.
- *Electrochemical deposition* is similar to electroplating of a metal layer [66].

7.4.2 Fabrication of Microfluidic Devices with Micro-Stereo Lithography

Because of its flexibility in making any three-dimensional structure, micro-stereo lithography can be used for the fabrication of complex microfluidic devices. The technique allows the fabrication of microvalves [67] and integrated fluid systems [68], which have a microchannel network embedded in the transparent polymeric device. Microreactors with embedded photodiodes for biochemical analysis have been reported by Ikuta et al. [69]. Active devices such as SMA-actuated micropumps and microvalves can be integrated in the above systems for controlling fluid flow. The technique of layer-by-layer photolithography is used for fabricating static Y-mixers [70]. The flexibility of the technique allows scaling down of the complex geometry of a macroscopic mixer. The advantage of micro-stereo lithography is the self-packaged approach. While fluidic interconnects are serious problems for microfluidic devices made with other microtechniques, interconnects are easily fabricated in the same process. Problems of leakage and expensive packaging can be solved with micro-stereo lithography. Different microfluidic manufacturing methods are summarized in Table 7.1.

Table 7.1 Summary of microfluidic channel manufacturing methods

Manufacturing methods	Principle	Materials	Features
Wet silicon etching	Chemical removal of layer	Silicon, silicon dioxide, silicon nitride	Low cost, high surface area, low aspect ratio
Dry silicon etching	Plasma-assisted etching	Silicon, silicon dioxide, silicon nitride	High surface area, minimum feature size, high aspect ratio
Lithography	Series of chemical reactions	Aluminum, steel, glass	Minimum feature size, maximum surface area, choice of geometry
Laser ablation	Bond-breakage by pulsed UV source	Copper, steel	High aspect ratio, choice of geometry, minimum feature size
LIGA	Uses X-ray or UV light	Silicon wafer, PMMA, SU-8	Minimum feature size, high aspect ratio, choice of geometry
Micromachining	Cutting of materials with tool	Steel, copper, aluminum	Maximum lifetime, choice of geometry, maximum surface area
μ-EDM	Melting and evaporation of material	Aluminum, copper	Maximum life time, high aspect ratio

7.5 Microfluidic Devices

Various components of a microfluidic device include microchannels, micoronozzles, micropumps, micromixers, and microvalves. These devices enable handling of miniscule volumes of fluid and can also be integrated into microfluidic chip or lab-on-a-chip.

7.6   Fluid Flow in Microfluidic Devices

Liquid flow can be laminar, transitional, or turbulent, in macroscopic scale, depending on the Reynolds number (R_e). The flow is laminar when the $R_e < 2100$, turbulent when $R_e > 4000$, and is partly laminar with intermittent bursts of irregular behavior when R_e is in the range of 2100–4000. At $R_e > 4000$, the flow is turbulent, accompanied by random fluctuations with particle mixing. The two components of velocity are unsteady components normal to the channel axis and the predominant component along the channel. In microscale devices having dimensions in the order of microns, the value of R_e is very small (<1), despite the velocity being high. The fluid shows laminar flow behavior, moving in smooth streamlines and rarely found turbulent. However, in this analysis of fluid behavior, only liquid flow is considered and the effect of compressibility is ignored. Thus, the flow is treated as in compressible flow. Compressible flow is characterized by constant density with respect to pressure and temperature. Though density is affected by temperature changes, the temperature is considered to be constant in this analysis. The flow behavior is further restricted to Newtonian fluid, particularly water, in which case the viscosity is constant.

7.7   Microchannels

As soon as the fluid enters a microchannel, it flows through the channel in two distinct regions of flow. At the entrance region, in the beginning, the flow profile significantly changes from flat to round and finally to the distinctive parabolic shape. Upon reaching this position where the fluid profile is parabolic, it flows in the fully developed region.

7.8   Applications of Microfluidics

7.8.1   Chemical Synthesis

A microchannel reactor is a device in which chemical reactions take place in a confinement with typical lateral dimensions below

1 mm; the most typical form of such confinement is microchannels. Microreactors are studied in the field of microprocess engineering, together with other devices (such as microheat exchangers) in which physical processes occur. The microreactor is usually a continuous-flow reactor in contrast to a batch reactor. Microreactors offer many advantages over conventional-scale reactors, including vast improvements in energy efficiency, reaction speed and yield, safety, reliability, scalability, on-site/on-demand and production, and a much finer degree of process control.

7.8.2 Separation and Analysis

Many chemical and biochemical analysis methods involve performing a sequence of processes that can be broadly classified in terms of sample preparation, reactions, and product analysis. Since the reaction products often contain mixtures of multiple chemical species, subsequent analytical steps must be capable of separating and identifying the individual components. Electrophoresis, which relies on inducing detectable differences in migration behavior between charged species under the influence of an applied electric field, has proven to be a highly versatile analytical technique owing to a favorable combination of characteristics, including relatively simple hardware design and compatibility with a wide range of analytes, including biological macromolecules.

7.8.3 Biodetection

Biodetection refers to the field of medical diagnostics, food quality, and biological warfare detection. The aim of microfluidic detection devices is to miniaturize and parallel classical immunologic and genomic detection assays. Micro- and nanofluidic devices dedicated to biodetection can be divided into two major classes: (i) sample preparation devices in which a preconditioning of the sample can be obtained (matrix change, pre-concentration, cell-lysis, purification, etc.) and (ii) biosensor devices in which the presence of the targeted analyte is transformed into an electrical or optical signal. On-chip real-time PCR, enzyme or classic ELISA immunosensors, and microarrays are among the most promising technologies for biological agent or marker detection.

7.8.4   Single Cell Biology

Microfluidics is a well-understood physics domain and can now be used to develop tools for cell biology. By simply miniaturizing macroscopic systems and taking advantage of the possibility of massive parallel processing, some microfluidic chips enable high-throughput biological experiments. Specific effects of laminar flow at the micron-scale also enable spatial control of liquid composition at subcellular resolution, fast media and temperature changes, and single cell handling and analysis. Microfluidic technology enables studies of cell behavior from single- to multicellular organism level with precise and localized application of experimental conditions unreachable using macroscopic tools.

7.8.5   Microdroplets

Multiphase flows generate a high interest in microfluidics as the laminar flows facilitate the generation of monodisperse droplets. Emulsions and double emulsions can be used for nanoparticles synthesis, drug microencapsulation (lipid vesicles), and active substance encapsulation. Microdroplets can also be used as single microreactors in biodetection systems. The amplification of single DNA strands can be obtained to increase the sensitivity of the biodetection schema.

Alternatives to the above closed-channel continuous-flow systems include novel open structures, where discrete, independently controllable droplets are manipulated on a substrate using electro wetting. Following the analogy of digital microelectronics, this approach is referred to as digital microfluidics.

7.8.6   Microfluidic Rheology/Rheometry

Working with fluids with low Reynolds numbers enables to properly investigate their viscosity, especially non-linear viscous effects. A number of microfluidic rheometry systems have been investigated, and they enable an alternative to conventional characterization methods. The study of fluid transport across micro-/nanofluidic porous media is also of high interest for oil recovery applications. Finally, a number of fundamental questions around the slip velocity at the solid–liquid interface can also be investigated using microfluidic devices together with particle velocimetry techniques.

7.8.7 Optofluidics

Optofluidics refers to the manipulation of light using fluids, or vice versa, on the micro- to nanometer scale. By taking advantage of the microfluidic manipulation, the optical properties of the fluids can be precisely and flexibly controlled to realize reconfigurable optical components, which are otherwise difficult or impossible to implement with solid state technology. In addition, the unique behavior of fluids on the micro-/nanoscale has given rise to the possibility to manipulate the fluid using light.

7.8.8 Drug Delivery

Drug delivery is defined as the administration of a chemical compound to any biological system for therapeutic purposes. Drug delivery can be achieved using conventional methods, including oral administration, inhalation, or injection through skin. However, one of the major drawbacks of these methods is that the drug's pathway (the physical distance between the inoculation and the target site) is quite long. This long pathway renders the treatment potentially ineffective. To overcome this drawback, microfluidics can offer many solutions and make possible more effective and targeted drug delivery. Drug delivery at the cellular level, tissue level, and organism level can be achieved. The lab-on-a-chip, one of microfluidics' main applications, can offer a platform for drug synthesis and delivery. Another example is microfluidic gradient generators (MGGs). The techniques on which these are based include time-evolving diffusion and parallel streams mixing. In the first technique, devices consist of two reservoirs having high and low concentrations of one or more reagents. A bridge having a built-in valve is used to connect these reservoirs, where cell culture commonly takes place. Figure 7.3 shows the schematic of a sink–source flow-free gradient generator. The key feature of the device is the absence of convection flow indicating no shear-stress induced to cells.

In the second type of MGGs, the microfluidic device consists of parallel streams flowing continuously within it to produce stable concentration gradient suitable for long-term cell observations. This can be achieved by regulating the diffusive mixing of the reactant streams at the interface of the flow. A very common type belonging to this category of microfluidic devices is tree-like gradient generators (TLGG). Figure 7.4 represents a schematic of TLGG.

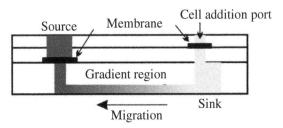

Figure 7.3 Schematic showing side view of a diffusion-based gradient generator. The *x*-direction of the channel contains chemical gradient [71].

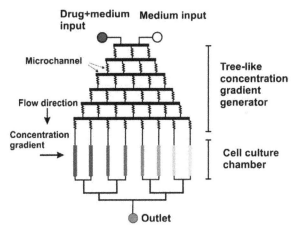

Figure 7.4 Details of the single structure unit showing an upstream concentration gradient generator and downstream cell culture chambers. The device consists of eight uniform structure units. Each unit is connected by a common reservoir in the center of the device [72].

For further examples of microfluidic devices for drug delivery, readers can follow Refs. [73–75].

The following are the advantages of microfluidic devices in drug delivery applications:

- Reduction in both pain and risk of side effects

- Further improvements in cost, user-friendliness, safety, and portability
- Precise and controlled drug delivery
- High resolution and real-time observations
- Concentrations of the drug to be administered tunable

7.8.9   Microfluidic Microneedles

To enhance the effectiveness of drug delivery and reduce pain, needles can be miniaturized. Microneedles are further divided into four types: (i) solid microneedles, (ii) microneedles coated with drug, (iii) polymeric microneedles encapsulating drug that can fully dissolve in the skin, and (iv) hollow microneedles for drug infusion into skin.

Microfluidic microneedles may contain hollow microneedle(s) singularly or in an array. A single microfluidic device containing an array of hollow microneedles provides the possibility of delivery of liquid formulations to larger areas, thereby allowing a faster intradermal delivery. This has been made possible by advanced micro-electromechanical system (MEMS) production techniques, allowing cost effectiveness and better performance.

Recently, the design and fabrication of a bulk titanium microneedle device composed of microneedle arrays have been carried out [76]. A controlled fluid delivery through the microfluidic network embedded within these arrays has been achieved. Polymeric microneedles have also been developed for transdermal drug delivery applications [77].

7.8.10   Microfluidic Micromixers

Miniaturization of fluid handling and fluid analysis is becoming important in analytical chemistry. Fast mixing is vital in most of the microfluidic systems that are used in biochemistry analysis, drug delivery, sequencing or synthesis of nucleic acids, other biological processes such as cell activation, enzyme reactions, and protein folding, which frequently involve reactions requiring mixing of reactants for the initiation of a reaction. Lab-on-a-chip platforms also require mixing for complex chemical reactions. Microscale mixers provide miniaturized systems designed specifically for applications, such as nanoparticle crystallization, extraction, polymerization, organic synthesis, enzyme assay, protein folding, biological screening, analytical assay, cell analysis, bioprocess optimization, clinical diagnostics, and drug delivery studies, for an intimate contact between the reactant molecules to facilitate chemical reactions. Micromixers can also be integrated in a microfluidic system [78].

In a microfluidic environment, enhanced control over fluid properties becomes apparent, which offers the likelihood of employing control over other processes carried out at the microscale level. Another advantage of microfluidic systems is that they provide an opportunity to spatially and temporally observe and control reactions by the addition of reagents at specific time intervals during the progress of a chemical reaction. The amount of sample required for a reaction is also reduced due to the small internal volume of the microfluidic mixers. This is mainly important when the reactions involve rare and expensive substances or samples, or a large number of samples in a restricted small volume. A typical channel size with a microfluidic device is 10–400 µm. Due to such small dimensions, an increase in the surface-to-volume ratio to 10,000–50,000 $m^2 m^{-3}$ compared with 100–2000 $m^2 m^{-3}$ of their macroscale counterparts. This characteristic feature of microfluidic channels brings about a superior heat transfer and control, facilitating reactions to take place in an isothermal manner. This feature is particularly useful in handling exothermal reactions within the micromixer where fast heating and cooling of the reaction mixture are required.

Further, highly monodisperse droplets can be generated in a parallelized manner by microfluidic devices, as required for many industrial applications, and evasion of cross-contamination between the droplets can be made possible by separating the droplets with an immiscible fluid or gas. The small internal volume of the micromixers is also advantageous for safety while handling toxic substances/chemicals and chemical reactions. This is exemplified by the possibility of generation of H_2O_2 (on demand and in situ), which is one of the most toxic substances [79].

7.9 Conclusion

Microfluidics deals with the manipulation of little quantity of fluids, using channels having dimensions of tens to hundreds of micrometers. The first applications of microfluidic technologies have been in analysis, for which they offer a number of useful capabilities: the ability to use very small quantities of samples and reagents and to carry out separations and detections with high resolution and sensitivity; low cost; short times for analysis; and small footprints for the analytical devices. Microfluidics exploits the most obvious characteristic of fluids in microchannels—small size—and the less obvious ones, such as laminar flow. It offers fundamentally new capabilities for controlling concentrations of molecules in space and time.

Problems

1. Differentiate between cross-flow and dead-end filtration techniques with examples.
2. Explain plume mixing techniques. Describe the role of filters in microfluidic devices.
3. Give the design sequence for lab-on-a-chip devices.
4. Explain the governing forces in microfluidics.
5. Describe the fabrication sequence for a bubble pump.

References

1. Lammertyn, J., Verboven, P., Veraverbeke, E. A., Vermeir, S., Irudayaraj, J., and Nicolaï, B. M. (2006). Analysis of fluid flow and reaction kinetics in a flow injection analysis biosensor, *Sensors and Actuators B: Chemical,* **114(2)**, pp. 728.

2. Kuswandi, B., Huskens, J., and Verboom, W. (2007). Optical sensing systems for microfluidic devices: A review, *Analytica Chimica Acta,* **601(2)**, pp. 141–155.

3. Gould, P. (2004). Microfluidics realizes potential, *Materials Today,* **7(7)**, pp. 48–52.

4. Woias, P. (2005). Micropumps: Past, progress and future prospects, *Sensors and Actuators B: Chemical,* **105**, pp. 28.

5. Stone, H. A., and Kim, S. (2001). Microfluidics: Basic issues, applications, and challenges, *AlChE Journal,* **47(6)**, pp. 1250.

6. Gervais, T., and Jensen, K. F. (2006). Mass transport and surface reactions in microfluidic systems, *Clinical Engineering Science,* **61**, pp. 1102.

7. Pepin, A., Youinou, P., Studer, V., Lebib, A., and Chen, Y. (2002). Nanoimprint lithography for the fabrication of DNA electrophoresis chips, *Microelectronic Engineering,* **61–62**, pp. 927–932.

8. Wu, C.-H., Chun-Hwa, C., Kuo-Wei, F., Wen-Syang, H., and Yu-Cheng, L. (2007). Design and fabrication of polymer microfluidic substrates using the optical disc process, *Sensors and Actuators A: Physical,* **139(1)**, pp. 310.

9. Ruano-Lopez, J. M., Aguirregabiria, M., Tijero, M., Arroyo, M. T., Elizalde, J., Berganzo, J., Aranburu, I., Blanco, F. J., and Mayora, K. (2006). A new SU-8 process to integrate buried waveguides and sealed microchannels for a Lab-on-a-Chip, *Sensors and Actuators B: Chemical,* **114(1)**, pp. 542.

10. Xiuhua, S., Peeni, B. A., Yang, W., Becerril, H. A., and Woolley, A. T. (2007). Rapid prototyping of poly(methyl methacrylate) microfluidic systems using solvent imprinting and bonding, *Journal of Chromatograph A,* **1162(2)**, pp. 162.

11. Yu, L., Tay, F. E. H., Xu, G., Chen, B., Avram, M., and Iliescu, C. (2006). Adhesive bonding with SU-8 at wafer level for microfluidic devices, *Journal of Physics: Conference Series,* **34(1)**, pp. 776.

12. Bilenberg, B., Nielsen, T., Clausen, B., and Kristensen, A. (2004). PMMA to SU-8 bonding for polymer based lab-on-a-chip systems with integrated optics, *Journal of Micromechanics and Microengineering,* **14(6)**, pp. 814.

13. Liu, J.-S., Liu, C., Guo, J.-H., and Wang, L.-D. (2006). Electrostatic bonding of a silicon master to a glass wafer for plastic microchannel fabrication. *Journal of Materials Processing Technology,* **178(1)**, pp. 278.

14. Poenar, D. P., Iliescu, C., Carp, M., Pang, A. J., and Leck, K. J. (2007). Glass-based microfluidic device fabricated by parylene wafer-to-wafer bonding for impedance spectroscopy, *Sensors Actuators A: Physical,* **139(1)**, pp. 162.

15. Capretto, L., Cheng, W., Hill, M., and Zhang, X. (2011). Micromixing within microfluidic devices, *Topics in Current Chemistry,* **304**, pp. 27.

16. Allen, P. B., and Chiu, D. T. (2008). Calcium-assisted glass-to-glass bonding for fabrication of glass microfluidic devices, *Analytical Chemistry,* **80**, pp. 7153–7157.

17. Vulto, P., Huesgen, T., Albrecht, B., and Urban, G. A. (2009). A full-wafer fabrication process for glass microfluidic chips with integrated electroplated electrodes by direct bonding of dry film resist, *Journal of Micromechanical Microengineering,* **19**, pp. 077001.

18. Li, P. C. H., and Harrison, D. J. (1997). Transport, manipulation, and reaction of biological cells on-chip using electrokinetic effects, *Analytical Chemistry,* **69**, pp. 1564–1568.

19. Waters, L. C., Jacobson, S. C., Kroutchinina, N., Khandurina, J., Foote, R. S., and Ramsey, J. M. (1998). Microchip device for cell lysis, multiplex PCR amplification, and electrophoretic sizing, *Analytical Chemistry,* **70**, pp. 158–162.

20. Ichiki, T., Ujiie, T., Shinbashi, S., Okuda, T., and Horiike, Y. (2002). Immuno-electrophoresis of red blood cells performed on microcapillary chips, *Electrophoresis,* **23**, pp. 2029–2034.

21. Lee, H., Sun, E., Ham, D., and Weissleder, R. (2008). Chip-NMR biosensor for detection and molecular analysis of cells, *Nature Medicine,* **14**, pp. 869–874.

22. Qu, B.-Y., Wu, Z.-Y., Fang, F., Bai, Z.-M., Yang, D.-Z., and Xu, S.-K. (2008). A glass microfluidic chip for continuous blood cell sorting by a magnetic gradient without labelling, *Analytical Bioanalytical Chemistry,* **392**, pp. 1317–1324.

23. Jacobson, S. C., Koutny, L. B., Hergenroder, R., Moore, A. W., and Ramsey, J. M. (1994). Microchip capillary electrophoresis with an integrated post-column reactor, *Analytical Chemistry,* **66**, pp. 3472–3476.

24. Liang, Z.-H, Chiem, N., Ocvirk, G., Tang, T., Fluri, K., and Harrison, D. J. (1996). Microfabrication of a planar absorbance and fluorescence cell for integrated capillary electrophoresis devices, *Analytical Chemistry,* **68**, pp. 1040–1046.

25. Fluri, K., Fitzpatrick, G., Chiem, N., and Harrison, D. J. (1996). Integrated capillary electrophoresis devices with an efficient post column reactor in planar quartz and glass chips, *Analytical Chemistry,* **68**, pp. 4285–4290.

26. Jacobson, S. C., Culbertson, C. T., Daler, J. E., and Ramsey, J. M. (1998). Microchip structures for sub-millisecond electrophoresis, *Analytical Chemistry,* **70**, pp. 3476–3480.

27. Ujiie, T., Ichiki, T. K., and Horiike, Y. (2000). Fabrication of quartz microcapillary electrophoresis chips using plasma etching, *Japanese Journal of Applied Physics,* **39**, pp. 3677–3682.

28. Lee, T. M. H., Hsing, I.-M., Lao, A. I. K., and Carles, M. C. (2000). A miniaturized DNA amplifier: Its application in traditional Chinese medicine, *Analytical Chemistry*, **72**, pp. 4242–4247.

29. Deng, Y.-Z., Zhang, H.-W., and Henion, J. (2001). Chip-based quantitative capillary electrophoresis/mass spectrometry determination of drugs in human plasma, *Analytical Chemistry*, **73**, pp. 1432–1439.

30. Gottschlich, N., Jacobson, S. C., Culbertson, C. T., and Ramsey, J. M. (2001). Two-dimensional electrochromatography/capillary electrophoresis on a microchip, *Analytical Chemistry*, **73**, pp. 2669–2674.

31. Omasu, F., Nakano, Y., and Ichiki, T. (2005). Measurement of the electrophoretic mobility of sheep erythrocytes using microcapillary chips, *Electrophoresis*, **26**, pp. 1163–1167.

32. Mukhopadhyay, R. (2000). When PDMS isn't the best, *Analytical Chemistry*, **79**, pp. 3248–3253.

33. Becker, H., and Heim, U. (2000). Hot embossing as a method for the fabrication of polymer high aspect ratio structures, *Sensors and Actuators A: Physical*, **83**, pp. 130–135.

34. Qi, S.-Z., Liu, X.-Z., Ford, S., Barrows, J., Thomas, G., Kelly, K., McCandless, A., Lian, K., Goettert, J., and Soper, S. A. (2002). Microfluidic devices fabricated in poly(methyl methacrylate) using hot embossing with integrated sampling capillary and fiberoptics for fluorescence detection, *Lab on a Chip*, **2**, pp. 88–95.

35. Chien, R.-D. (2006). Micromolding of biochip devices designed with microchannels, *Sensors and Actuators A: Physical*, **128**, pp. 238–247.

36. Attia, U. M., Marson, S., and Alcock, J. R. (2009). Micro-injection moulding of polymer microfluidic devices, *Microfluidics and Nanofluidics*, **7**, pp. 1–28.

37. McDonald, J. C., Duffy, D. C., Anderson, J. R., Chiu, D. T., Wu, H.-K., Schueller, O. J. A., and Whitesides, G. M. (1999). Fabrication of micro-fluidic systems in poly(dimethylsiloxane), *Electrophoresis*, **21**, pp. 27–40.

38. Rossier, J., Reymond, F., and Michel, P. E. (2002). Polymer microfluidic chips for electrochemical and biochemical analyses, *Electrophoresis*, **23**, pp. 858–867.

39. Cao, H., Tegenfeldt, J.O., Austin, R. H., and Chou, S. Y. (2002). Gradient nanostructures for interfacing microfluidics and nanofluidics, *Applied Physics Letters*, **81**, pp. 3058–3060.

40. Mappes, T., Achenbach, S., and Mohr, J. (2007). X-ray lithography for devices with high aspect ratio polymer submicron structures, *Microelectronics Engineering,* **84**, pp. 1235–1239.

41. Odom, T. W., Love, J. C., Wolfe, D. B., Paul, K. E., and Whitesides, G. M. (2002). Improved pattern transferins of lithography using composite stamps, *Langmuir,* **18**, pp. 5314–5320.

42. Hua, F., Sun, Y.-G., Gaur, A., Meitl, M. A., Bilhaut, L., Rotkina, L., Wang, J.-F., Geil, P., Shim, M., and Rogers, J. A. (2004). Polymer imprint lithography with molecular scale resolution, *Nano Letters,* **4**, pp. 2467–2471.

43. Xu, Q.-B., Mayers, B. T., Lahav, M., Vezenov, D. V., and Whitesides, G. M. (2005). Approaching zero: Using fractured crystals in metrology for replica molding, *Journal of American Chemical Society,* **127**, pp. 854–855.

44. Li, Z.-W., Gu, Y.-N., Wang, L., Ge, H.-X., Wu, W., Xia, Q.-F., Yuan, C.-S., Cheng, Y.-F., Cui, B., and Williams, R. S. (2009). Hybrid nanoimprints of lithography with sub- 15 nm resolution. *Nano Letters,* **9**, pp. 2306–2310.

45. Bruzzone, A. A. G., Costa, H. L., Lonardo, P. M., and Lucca, D. A. (2008). Advances in engineered surfaces for functional performance, *CIRP Annals-Manufacturing Technology,* **57**, pp. 750.

46. Ungerbock, B., and Mistlberger, G. (2010). Oxygen imaging in microfluidic devices with optical sensors applying color cameras, *Procedia Engineering,* **5**, pp. 456.

47. Drăgoi, V., Cakmak, E., and Pabo, E. (2010). Metal wafer bonding for MEMS devices, *Romanian Journal of Information Science And Technology,* **13(1)**, pp. 65.

48. Ackermann, K. R., Henkel, T., and Popp, J. (2007). Quantitative online detection of low concentrated drugs via a SERS microfluidic system, *Chem Phys Chem,* **8**, pp. 2665.

49. Becker, E. W., Ehrfeld, W., Hagmann, P., Maner, A., and Münchmeyer, D. (1986). Fabrication of microstructures with high aspect ratios and great structural heights by synchrotron radiation lithography, glavanoforming, and plastic moulding (LIGA process), *Microelectronic Engineering,* **4**, pp. 35–56.

50. Guckel, H., Christensen, T. R., and Skrobis, K. J. (1995). Formation of microstructures using a preformed photoresist sheet, U.S. Patent #5378583, January 1995.

51. Chaudhuri, B., Guckel, H., Klein, J., and Fischer, K. (1998). Photoresist application for the LIGA process, *Microsystem Technologies,* **4**, pp. 159–162.

52. Mohr, J., Ehrfeld, W., and Münchmeyer, D. (1988). Requirements on resist layers in deep-etch synchrotron radiation lithography, *Journal of Vacuum Science and Technology*, **B6**, pp. 2264–2267.

53. Guckel, H., et al., "Plasma Polymerization of Methyl Methacrylate: A Photoresist for 3D Applications," Technical Digest of the IEEE Solid State Sensor and Actuator Workshop, Hilton Head Island, SC, June 4–7, 1988, pp. 43–46.

54. Ghica, V., and Glashauser, W. (1982). Verfahrenfr die Spannung sfreie Entwicklung von Bestrahlten Polymethylmethacrylate- Schichten, German patent, #3039110, 1982.

55. Lee, K. Y., LaBianca, N., Rishton, S. A., Zolgharnain, S., Gelorme, J. D., Shaw, J., and Chang, T. P. (1995). Micromachining applications for a high resolution ultra-thick photoresist, *Journal of Vacuum Science and Technology B*, **13**, pp. 3012–3016.

56. Shaw, J. M., Gelorme, J. D., LaBianca, N. C., Conley, W. E., and Holmes, S. J. (1997). Negative photoresists for optical lithography, *IBM Journal of Research and Development*, **41**, pp. 81–94.

57. Lorenz, H., Despont, M., Fahrni, N., LaBianca, N., Renaud, P., and Vettiger, P. (1997). SU-8: A low-cost negative resist for MEMS, *Journal of Micromechanics and Microengineering*, **7**, pp. 121–124.

58. Lorenz, H., Despont, M., Vettiger, P., and Renaud, P. (1998). Fabrication of photoplastic high-aspect ratio microparts and micromolds using SU-8 UV resist, *Microsystem Technologies*, **4**, pp. 143–146.

59. Conédéra, V., Le Goff, B., and Fabre, N. (1999). Potentialities of a new positive photoresist for the realization of thick moulds, *Journal of Micromechanics and Microengineering*, **9**, pp. 173–175.

60. Loechel, B. (2000). Thick-layer resists for surface micromachining, *Journal of Micromechanics and Microengineering*, **10**, pp. 108–115.

61. Frazier, A. B., and Allen, M. G. (1993). Metallic microstructures fabricated using photosensitive polyimide electroplating molds, *Journal of Microelectromechanical Systems*, **2(2)**, pp. 87–94.

62. Ito, T., Sawada, R., Higurashi, E., and Kiyokura, T. (2000). Fabrication of microstructure using fluorinated polyimide and silicon-based positive photoresist, *Microsystem Technologies*, **6**, pp. 165–168.

63. Stieglitz, T. (2001). Flexible biomedical microdevices with double-sided electrode arrangements for neural applications, *Sensors and Actuators A: Physical*, **90**, pp. 203–211.

64. Metz, S., Holzer, R., and Renaud, P. (2001). Polyimide-based microfluidic devices, *Lab on a Chip*, **1(1)**, pp. 29–34.

65. O'Brien, J., Hughes, P. J., Brunet, M., O'Neill, B., Alderman, J., Lane, B., O'Riordan, A., and O'Driscoll, C. (2001). Advanced photoresist technologies for microsystems, *Journal of Micromechanics and Microengineering*, **11**, 2001, pp. 353–358.

66. Smela, E. (1998). Microfabrication of PPy microactuators and other conjugated polymer devices, *Journal of Micromechanics and Microengineering*, **9**, pp. 1–18.

67. Ikuta, K., and Hirowatari, K. (1993). Real three dimensional micro fabrication using stereo lithography and metal molding. In: *Proceedings of MEMS'93, 6th IEEE International Workshop on Micro Electromechanical System*, January 25–28, 1993, San Diego, California, USA, pp. 42–47, IEEE.

68. Ikuta, K., Hirowatari, K., and Ogata, T. (1994). Three dimensional micro integrated fluid systems (MIFS) fabricated by stereo lithography. In: *Proceedings of MEMS'94, 7th IEEE International Workshop on Micro Electromechanical System*, January 25–28, 1994, Oiso, Japan, pp. 1–6, IEEE.

69. Ikuta, K., Maruo, S., Fukaya, Y., and Fujisawa, T. (1998). Biochemical IC chip toward cell free DNA protein synthesis. In: *Proceedings of MEMS'98, 11th IEEE International Workshop on Micro Electromechanical System*, January 25–29, 1998, Heidelberg, Germany, pp. 131–136, IEEE.

70. Bertsch, A., Heimgartner, S., Cousseau, P., and Renaud, P. (2001). Static micromixers based on large-scale industrial mixer geometry, *Lab on a Chip*, **1**, pp. 56–20.

71. Abhyankar, V. V., Lokuta, M. A., Huttenlocher, A., and Beebe, D. J. (2006). Characterization of a membrane-based gradient generator for use in cell-signaling studies, *Lab on as Chip*, **6**, pp. 389–393.

72. Ye, N., Qin, J., Shi, W., Liu, X., and Lin, B. (2007). Cell-based high content screening using an integrated microfluidic device, *Lab on a Chip*, **7**, pp. 1696–1704.

73. Adams, T., Minerick, A. R., Yang, C., Gress, J., and Wimmer, N. (2012). A tunable microfluidic device for drug delivery. In: *Advances in Microfluidics*, Kelly, R. (ed.), InTech. DOI: 10.5772/37355. Available at: www.intechopen.com/books/advances-in-microfluidics/exploring-a-digital-microfluidic-device-for-gastric-cancer-drug-delivery

74. Haeberle, S., Hradetzky, D., Schumacher, A., Vosseler, M., Messner, S., and Zengerle, R. (2009). Microfluidics for drug delivery. In: *World Congress on Medical Physics and Biomedical Engineering*, September 7–12, 2009, Munich, Germany, pp. 359–362, Springer Berlin Heidelberg.

75. Riahi, R., Tamayol, A., Shaegh, S. A. M., Ghaemmaghami, A. M., Dokmeci, M. R., and Khademshosseini, A. (2015). Microfluidics for advanced drug delivery systems, *Current Opinion in Chemical Engineering*, **7**, pp. 101–112.

76. Parker, E. R., Rao, M. P., Turner, K. L., and MacDonald, N. C. (2006). Bulk titanium microneedles with embedded microfluidic networks for transdermal drug delivery. In: *19th IEEE International Conference on Micro Electro Mechanical Systems*, January 22–26, 2006, Istanbul, Turkey, pp. 498–501, IEEE.

77. Bodhale, D. W., Nisar, A., and Afzulpurkar, N. (2010). Structural and microfluidic analysis of hollow side-open polymeric microneedles for transdermal drug delivery applications, *Microfluidics and Nanofluidics*, **8(3)**, pp. 373–392.

78. Nguyen, N.-T., and Wu, Z. (2005). Micromixers: A review, *Journal of Micromechanics and Microengineering*, **15**, pp. R1–R16.

79. Capretto, L., Cheng, W., Hill, M., and Zhang, X. (2011). Micromixing within microfluidic devices, *Topics in Current Chemistry*, **304**, pp. 27–68.

Index